POWER INVESTING with BASKET SECURITIES

THE INVESTOR'S GUIDE TO EXCHANGE-TRADED FUNDS

PETER W. MADLEM, CFA
LARRY D. EDWARDS, CFA

POWER INVESTING with BASKET SECURITIES

THE INVESTOR'S GUIDE TO EXCHANGE-TRADED FUNDS

ST. LUCIE PRESS

A CRC Press Company
Boca Raton   London   New York   Washington, D.C.

Library of Congress Cataloging-in-Publication Data

Madlem, Peter.
    Power investing with basket securities : the investor's guide to exchange-traded funds / Peter W. Madlem, Larry D. Edwards.
    p. cm.
    Includes bibliographical references and index.
    ISBN 1-57444-254-6
    1. Exchange traded funds. I. Edwards, Larry D. II. Title.

HG6043 .M33 2001
332.63′228—dc21   2001048547

This book contains information obtained from authentic and highly regarded sources. Reprinted material is quoted with permission, and sources are indicated. A wide variety of references are listed. Reasonable efforts have been made to publish reliable data and information, but the author and the publisher cannot assume responsibility for the validity of all materials or for the consequences of their use.

Neither this book nor any part may be reproduced or transmitted in any form or by any means, electronic or mechanical, including photocopying, microfilming, and recording, or by any information storage or retrieval system, without prior permission in writing from the publisher.

The consent of CRC Press LLC does not extend to copying for general distribution, for promotion, for creating new works, or for resale. Specific permission must be obtained in writing from CRC Press LLC for such copying.

Direct all inquiries to CRC Press LLC, 2000 N.W. Corporate Blvd., Boca Raton, Florida 33431.

Trademark Notice: Product or corporate names may be trademarks or registered trademarks, and are used only for identification and explanation, without intent to infringe.

Visit the CRC Press Web site at www.crcpress.com

© 2002 by CRC Press LLC
St. Lucie Press is an imprint of CRC Press LLC

No claim to original U.S. Government works
International Standard Book Number 1-57444-254-6
Library of Congress Card Number 2001048547
Printed in the United States of America 1 2 3 4 5 6 7 8 9 0
Printed on acid-free paper

Preface

Power Investing with Basket Securities, The Investor's Guide to Exchange-Traded Funds, represents the second volume in the Power Investing series of books. Several years ago, when the publisher originally suggested the idea, there was some hesitancy regarding the series title. *Power* seemed, well, a bit robust as a title theme. However, upon consulting the dictionary, the primary definition was found to be, "the ability or capacity to act or perform effectively." Who can quarrel with that? After all, that is exactly what the series is intended to do, impart to readers the ability and capacity to perform effectively with their investments. Each volume presents an investment strategy or approach that empowers the investor, whether professional or individual, with proven and innovative approaches to improve performance in the convoluted and confounding world that makes up the financial marketplace.

This book is about basket securities and how investors can use them. Spiders, Sectors, Holders, ETFs, iShares, Vipers, and Qubes, are all acronyms for members of this relatively new class of financial products. Most investors know them as exchange-traded funds, or ETFs. However, not all basket securities are exchange-traded funds, in the strictest sense. The fundamental concept of a basket is any grouping or combination of financial assets that underlie the value of a single exchange-traded security. Many, but not all, are based on indices, or subsectors of indices. In addition, most are structured as securities issued by unit investment trusts or mutual funds, but not all. The term "basket security" covers a broader spectrum of useful securities than just index fund products issued by registered investment companies.

Most baskets are designed to track specific indices or subgroups. Others select specified portfolios within an industry group such as telecommunications and pharmaceuticals. Because they are basically passive portfolios, baskets incur lower expenses than actively managed mutual funds. Moreover, they typically entail a lesser tax burden, since the underlying stocks are generally traded when an index needs to be rebalanced.

Like stocks, you can buy baskets on margin and trade them anytime the market is open. Baskets also are easy to short because they are exempt from the uptick rule, which forbids the selling of a stock short in a declining market. Baskets can be used to manage portfolio risk through hedging. Professional traders can swap shares of a basket security for the underlying stocks, or vice versa. This form of arbitrage keeps basket security prices very close to their net asset value, which means they can track an index better than a closed-end mutual fund.

This book describes the origins of basket securities and reviews their beneficial features and structures. It covers the broad array of currently available basket securities, and suggests the likelihood of others on the horizon. Most importantly, the book presents some strategies for their successful and powerful applications in managing portfolios.

Those who can benefit from reading and using this book include investors who have made asset allocation decisions as to their overall financial assets, and are now considering how to achieve an appropriate level of diversification in the equity portion of their portfolios. The authors hope that this book will be useful to its readers in accomplishing such diversification, and perhaps in providing some further insights into the portfolio management process.

About the Authors

Both authors are currently portfolio managers with the Trust Investment Services Division of Santa Barbara Bank & Trust, located in Santa Barbara, CA. Their professional backgrounds, while significantly different, share much in common.

Peter W. Madlem, CFA, has nearly 20 years of diverse investment experience. In the mid-1980s he cofounded two publishing and investment advisory firms specializing in closed-end funds and REITs. His market and financial comments have appeared in *The Wall Street Journal, Investor's Business Daily, Kiplinger's,* and *Money Magazine.*

A published author and composer, Madlem has written six investment-related books and his musical compositions have been performed at the Sydney Opera House, the National Cathedral in Washington, D.C., and London's Wigmore Hall, and have been recorded on Sony/Classical and Sonora records. Madlem earned a Master's degree from the University of California, Santa Barbara. He lives with his wife, Katherine, and two children in Carpinteria, CA.

Larry D. Edwards, CFA, has more than 30 years of broad experience in the investment management industry. In addition to being an experienced portfolio manager, he served as President of Leland O'Brien Rubinstein Associates Incorporated (LOR) and its subsidiary, SuperShare Services Corporation from 1984 to 1988. In these capacities, Edwards had an integral role in the creation of "The SuperTrust" and its various securities that are described in Chapter 1 and Appendix B. One of these securities, the Index Trust SuperUnit, was the first successful U.S. exchange-traded basket security, preceding the introduction of the Spider by about 3 months.

Prior to joining LOR as president, Edwards was chief investment officer for Western Asset Management Company. He began his professional career with Scudder Stevens and Clark Inc.

A graduate of Occidental College and the Stanford Graduate School of Business, Edwards and his wife, Janice, reside in the Santa Ynez Valley.

At Santa Barbara Bank & Trust, Madlem and Edwards, together with their associates, are developing creative applications for basket securities, some of which are described in this book.

Introduction

Tremendously versatile, basket securities have the potential to change the way money is managed. They are similar to index funds but trade like stocks, so they enjoy the advantages of both worlds. Baskets provide a way to obtain the diversification of, say, the Standard & Poor's 500 Stock Index in a single security that can be bought and sold during market hours just like any other stock.

Unlike regular mutual fund shares, which calculate their net asset values at the end of each trading day, baskets trade continually on a stock exchange, where their prices can change from trade to trade.

This book provides the proper definition of the basket security, a brief exploration of their true history, and simple yet powerful ways to exploit their advantages. To that end the book is organized into three broad sections containing nine chapters, and three appendices.

Section I contains Chapters 1 through 3. The first chapter reviews the origin of basket securities, as we know them today. The need for baskets became clear during the stock market "Crash" of October 1987. It took 5 years to develop one that worked and that had a sound regulatory structure.

In Chapter 2, some of the primary structural characteristics and related benefits of baskets are reviewed, using the "Spider," or SPDR, as a model. These innovative securities can provide instant diversification, price efficiency, tax efficiency, liquidity, and cost efficiency. When used as part of equity portfolios, they can reduce the likelihood of bombshells and, over time, assure freshness.

Chapter 3 surveys the major participants in the basket securities business and those firms that are sponsors or managers. Many of the currently available baskets have come to market only recently.

Section II, encompassing Chapters 4 through 7, represents the applied portion of the book. Chapter 4 explores simple, yet effective, ways basket securities can be used in asset management strategies, including trading the market, building a diversified core, or creating a "thousand stock portfolio."

Chapter 5 presents a more technical look at trading strategies using basket securities, including using baskets together with individual securities, the application of momentum screens, and how to utilize straightforward and useful charting techniques. A sample basket portfolio strategy, the Basket Case Portfolio, is introduced.

Chapter 6 acknowledges the ongoing debate between active and passive management and suggests a portfolio management approach to blending them. Based on a comparison between the capitalization-weighted and equal-weighted S&P 500 returns, two variants are introduced to ascertain whether active or passive investment is more likely to succeed. Next, a method for determining the appropriate blend of the diversified core indices using an iterative historical value at risk computation is presented.

Chapter 7 explores the use of S&P 500 Select Sector SPDRs as the investment vehicle for the active portion of the Basket Case Portfolio. A sector approach is presented whereby an assessment is made regarding the current position in the business cycle in order to create a rationale for the selection of specific SPDRs for investment. Computation of relative strength is discussed with implications for verifying the market's perception of the business cycle position.

Section III consists of Chapters 8 and 9. Chapter 8 delves briefly into what is currently available through basket securities on the international scene, presents evidence on the efficacy of international diversification, and offers a momentum strategy for trading these securities.

Finally, Chapter 9 considers what may lie ahead in the ongoing development of basket securities and similar tools for portfolio management. Bonds and folios are reviewed.

Section IV contains the appendices. Appendix B describes the structure, securities, and investment payoffs of "The SuperTrust," a pioneering effort by a small company that, among other things, created the first U.S. exchange-traded basket security as we know them today. Appendix C describes the active/passive indicator.

Appendix A contains information and data on many exchange-traded basket securities and is presented as a convenient resource to readers.

Contents

SECTION I Origins and Participants

Chapter 1 Origins of Basket Securities 3

Prologue 3
The Need — Born of the Crash 3
    Early Failures 6
    An Early Success — The Index Trust SuperUnit 6
Along Came a Spider 7

Chapter 2 Structure and Benefits of Basket Securities 9

What Are They? 9
How Baskets Are Structured 9
Characteristics and Benefits of the Standard & Poor's
  Depositary Receipt (SPDR) 9
    Portfolio Deposits and Redemptions 10
    Exchange Trading 10
    Cost Efficiency 11
    Freshness 11
    Reducing Bombshells 11

Chapter 3 A Survey of Major Participants 13

The American Stock Exchange (AMEX) 13
Barclays Global Investors, N.A. (BGI) 14
The New York Stock Exchange (NYSE) 16
Merrill Lynch, Pierce, Fenner & Smith Incorporated (Merrill Lynch) 16

SECTION II Investment Strategies and Examples

Chapter 4 Asset Management Strategies 21

The Investment Process Generally 21
Uses of Basket Securities in Equity Portfolios 21
    Trading the Market 21
    As an Equity Portfolio 21
    Within an Equity Portfolio 22
    The Thousand Stock Portfolio 22

Chapter 5 An Example: Thousand Stock Portfolio 29
Allocating Assets 67
Five-Stage System 67
Investment Tools 68

Chapter 6 The Diversified Core 69
Step One — Determining the Allocation to the Diversified Core 72
    The BCP Active/Passive Indicator 72
Step Two — Determining the Diversified Core Blend 74

Chapter 7 The Active Portion 83
The S&P 500 Sector SPDRs 83
Step One — Examining the Business Cycle 84
    Consumer Expectations 91
    Industrial Production 91
    Inflation 91
    Interest Rates 91
    Yield Curve 92
    Stock Prices 92
Step Two — Sector Analysis 92
    Sector Rotation 92
    Early Expansion 94
    Middle Expansion 95
    Late Expansion 96
    Early Contraction 97
    Late Contraction 99
Step Three — Technical Analysis 100
Conclusion 104

SECTION III International Investing and Looking Forward

Chapter 8 Exchange-Traded Funds on International Markets 107
Reassessing International Investing 108
A Momentum Strategy for International Equity Markets 110

Chapter 9 The Future 111
Fixed-Income Baskets 111
Proliferation of Other Equity Products 112
Use of Trading Strategies 113

SECTION IV *Appendices*

Appendix A — Fund Pages 117
Appendix B — The SuperTrust 221
Appendix C — The Active/Passive Indicator 227

Index 255

Section I

Origins and Participants

1 Origins of Basket Securities

PROLOGUE

OCTOBER, 1987

> *From the close of trading on Tuesday, October 13, 1987, to the close of trading on October 19, 1987, the Dow Jones Industrial Average ("Dow") fell 769 points or 31 percent (see Figure). In those four days of trading, the value of all outstanding U.S. stocks decreased by almost $1.0 trillion. On October 19, 1987, alone, the Dow fell by 508 points or 22.6 percent. Since the early 1920's, only the drop of 12.8 percent in the Dow on October 28, 1929 and the fall of 11.7 percent the following day, which together constituted the Crash of 1929, have approached the October 19 decline in magnitude.*

Report of THE PRESIDENTIAL TASK FORCE on MARKET MECHANISMS, Nicholas F. Brady, Chairman; U.S. Government Printing Office, Washington, D.C., January 1988, page 1.

THE NEED — BORN OF THE CRASH

By the time of the stockmarket "Crash" in October of 1987, many institutional investors had concluded that they would frequently prefer to trade **portfolios** rather than individual stocks. Institutional investors such as the managers of corporate and public pension funds, profit-sharing funds, mutual funds, endowment funds, and foundations often had a need to control the overall equity exposure of the funds they managed. In addition, many large investment firms were actively engaged in "index arbitrage." Both of these interests involved the trading of market exposure.

Let's take a quick look at how certain trading practices contributed to the development of basket securities.

In 1987, the primary means of transacting in portfolios, or large amounts of equity market exposure at one point in time, were stock index futures contracts and program trading. A stock index futures contract is essentially an agreed upon obligation of the buyer to pay to the seller the difference in price of some index if the price declines, and of the seller to pay to the buyer the price difference if the price rises subsequent to entering into the agreement. Buying or selling futures contracts on an index such as the Standard & Poor's 500 Stock Index (S&P 500) was (and is) a way to effectively and efficiently buy or sell a substantial amount of equity exposure.

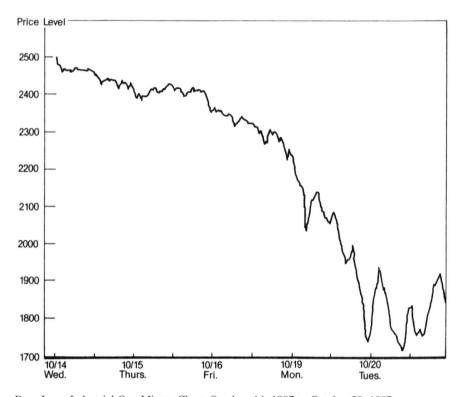

Dow Jones Industrial One-Minute Chart, October, 14, 1987 to October 20, 1987.

In October 1987 one S&P 500 Stock Index Futures contract had an underlying value of about $125,000 (i.e., 500 times the index level of about 250). Changes in this value were based upon (or derived from) changes in the price level of the index — hence, the term *derivative* instrument as applied to the futures contract. Trading volume of the S&P 500 contract was running at an annual rate of more than 20 million contracts, or about $2.5 trillion, roughly equal to the total dollar volume of trading on the U.S. stock market at that time. (Page II-13, Report of the Presidential Task Force on Market Mechanisms, January 1988.)

Program trades, by contrast, are the simultaneous execution of actual buy or sell orders for many stocks, often an entire index, such as the S&P 500. In 1987, program trades could be executed directly through the various specialist posts on the floors of the stock exchanges, or through the use of the designated order turnaround (DOT) automated execution system on the New York Stock Exchange.

Two primary types of program trades were (and are) used. Straight program trading takes place when there are no offsetting transactions in stock index futures contracts. Such a program trade, involving a basket or index of stocks, is typically used because of the speed and efficiency of execution, lower commission costs, and reduced market impact.

By contrast, a program trade executed simultaneously with an offsetting transaction in stock index futures contracts is known as index arbitrage. An index

Origins of Basket Securities

arbitrageur attempts to profit from perceived mispricing of the futures contract relative to the level of the index. (The difference in price of a futures contract and its underlying index is called the *basis*; usually the contract's price is higher than the index.) A futures price is considered abnormal when the basis differs from fair value, which relates to the difference between the yield on Treasury Bills and the dividend yield on the index, each calculated to the date of expiration of the futures contract. (Think of it as "cost of carry.") When an abnormal basis becomes large enough, the arbitrageur attempts to lock in a profit by simultaneously buying stocks and selling futures if the contract's price is too high, and by doing the reverse if the price of the futures is too low.

Index arbitrage acts to make the market more efficient. For example, if investors have bought or sold equity exposure through buying or selling stock index futures contracts, and in the process have created a futures price that is too high or too low, the subsequent action of the index arbitrageur tends to offset the mispricing and move the price of the futures back toward fair value. As the basis moves back toward fair value, the arbitrageur makes a profit, and the price relationship becomes more efficient.

The relevance of this quick review of futures, program trading, and index arbitrage to the 1987 stockmarket crash and the genesis of basket securities is that at that time the specialists at their posts had no way of knowing the source or purpose of the flood of sell orders they were receiving. Were they information-based trades from investors acting on new information? Were they part of straight program trades from investors managing the equity exposure of their portfolios? Or were they part of index arbitrage programs, taken due to perceived mispricing between futures and the index level?

The uncertainty as to the source and purpose of the multitude of sell orders arriving at their posts made the specialists more risk averse than would otherwise have been the case and less willing to commit capital to maintain an orderly market. And the volume of index arbitrage trading was insufficient to correct the abnormally large discount of the futures price relative to the index, in part due to the difficulty of executing program trades.

Following the crash, there were a number of studies published that attempted to diagnose its causes and prescribe remedies so that such an event would not be repeated. One such study, "The October 1987 Market Break: A Report by the Division of Market Regulation, U.S. Securities and Exchange Commission," suggested the following as one such remedial step:

> One of several alternatives that may be worthy of examination is the proposal to create one NYSE specialist post where the actual **market baskets** could be traded. A market basket post would alter the dynamics of program trading, in effect consolidating program trades back to a single order. The index specialist would have the informational advantage, not available to specialists in the individual stocks, of seeing the entire program order.... While the feasibility and design of **basket trading** would require substantial analysis, we believe the concept of basket trading deserves the Commission's and the NYSE's attention. (emphasis added) (*Black Monday and the Future of Financial Markets*, p. 366, 1989).

EARLY FAILURES

During the period following the October 1987 stockmarket crash, there were several attempts to develop viable basket securities. In May 1989, two exchange-created instruments began trading. One was the American Stock Exchange's equity index participations (EIPs); the other was the Philadelphia Stock Exchange's cash index participations (CIPs). Both began trading under Securities and Exchange Commission (SEC) registrations. The new securities featured procedures whereby sellers would deliver to buyers (or vice versa) the subsequent total return on the S&P 500 at some future time. The Chicago Mercantile Exchange and the Chicago Board Options Exchange promptly sued, claiming that the products were inherently futures contracts and should be regulated by the Commodity Futures Trading Commission (CFTC) rather than by the SEC. In August 1989, the U.S. Court of Appeals of the Seventh Circuit ruled that the products in fact should fall under CFTC jurisdiction. This ruling made it necessary for all investors to close their positions in the new Index Participations, and the instruments were delisted. (Harvard Business School Case N9-294-050; Leland O'Brien Rubinstein Associates, Inc.: SuperTrust™, page 14, June 6, 1994.)

AN EARLY SUCCESS — THE INDEX TRUST SUPERUNIT

Meanwhile, the small Los Angeles-based investment management firm of Leland O'Brien Rubinstein Associates Incorporated (LOR) was busily pursuing the creation of an innovative mutual fund/unit investment trust product called, "The SuperTrust." LOR had seen the demand soar for its Dynamic Asset Allocation (DAA) hedging strategies prior to the crash. Afterward, such demand largely evaporated. Indeed, various studies of the crash pointed to the selling by *portfolio insurers* (the name assigned LOR and other firms using dynamic hedging technology to implement protective strategies) as having at least exacerbated the severity of the decline.

Having witnessed heavy demand for investment payoff patterns that simultaneously featured downside protection together with upside capture, LOR was convinced that a substantial demand for such strategies could be rekindled if the delivery of the strategy could be assured. The means for such assurance would be to have the payoff backed with actual securities in the context of investment companies registered under The Investment Company Act of 1940 (the 1940 Act), the federal statute under which all mutual funds and unit investment trusts are regulated.

LOR established SuperShare Services Corporation (SSC) as a majority-owned subsidiary to create and bring the new product to market. (A further description of The SuperTrust is available in Appendix B.) For 5 years, from October 1987 until November 1992, SSC worked to design, develop, and market the SuperTrust product. Among the obstacles that were overcome were design features that were not previously accommodated by the 1940 Act, for which exemptive relief had to be obtained from the SEC.

Finally, on November 5, 1992, having raised approximately $1 billion from institutions and individuals as the initial investors, the SuperTrust was launched. One of the six securities was the Index Trust SuperUnit, a security that was issued by a unit investment trust under the 1940 Act, that was fully backed by the securities of an S&P 500 Index Fund managed by Wells Fargo Investment Advisors, and that traded on the American Stock Exchange. The first successful U.S. exchange-traded basket security had arrived!

In a Stock Index Research report dated November 1992, Goldman Sachs published a report titled, "Special Feature: *SuperTrust Is Launched!*" It included the following under "Who Are the Likely Users of SuperTrust?"

"We believe that the major users of the product will be the following:

- **Individual Investors** — For individual investors, the product provides a way to easily buy and sell the market in small amounts without the paperwork or operational hurdles of trading futures and options.
- **Active Investment Advisors** — Many active equity managers both inside and outside the U.S. do not have approval or are not set up operationally to handle futures. For them, the Index SuperUnit or SuperShare may be the easiest way to put cash into the market if they have not decided what specific stocks to buy or hedge market risk should they become bearish. Mutual funds may find them especially useful in helping them manage their cash balances.
- **Small Indexers** — Investment advisors wishing to establish small index funds for their clients may find SuperUnits a more efficient way to get index exposure than setting up their own fund."

The SuperTrust was on its way!

ALONG CAME A SPIDER

As described previously with regard to index participations, The American Stock Exchange (AMEX) had maintained a long-standing interest in the creation of basket securities. With two of the SuperTrust's securities trading on the AMEX (the other being the Money Market Trust SuperUnit), the AMEX had, of course, known about the SuperTrust in detail and had not only followed its developmental progress closely, but had also assisted in gaining clearance for exchange listings. Once the SuperTrust had received critical exemptive relief under the 1940 Act from the SEC, such relief could be (and was) conveniently cited by AMEX as precedent, thus vastly shortening the time for regulatory approval of similar products. In January 1993, 3 months after the SuperTrust launch, the AMEX dangled rubber spiders from the ceiling of its exchange, and to the delight of its members introduced the Standard & Poor's Depositary Receipt (SPDR, or Spider, exchange symbol "SPY"). The rest, as they say, is history, and is what the remainder of this book is about.

2 Structure and Benefits of Basket Securities

WHAT ARE THEY?

A basket security, as the term is used in this book, is a single security that trades on a stock exchange, that represents and tracks an underlying combination of securities (such as an index), and that has some structural mechanism to assure that its price on the exchange closely follows the value of its underlying securities. A primary benefit of such a security is that it can provide **instant diversification**. For example, in a single security an investor can own the risk and return characteristics of the entire S&P 500.

HOW BASKETS ARE STRUCTURED

Originally, basket securities were issued by unit investment trusts (UITs), which are investment companies regulated by the SEC under the Investment Company Act of 1940. UITs are similar to mutual funds, which are also registered investment companies under the 1940 Act. One significant difference is that a UIT has no board of directors and so is prevented by the 1940 Act from holding a managed portfolio. However, the SEC has permitted UITs to hold indexed portfolios that are only administered in accordance with preestablished guidelines.

While the structure and characteristics vary somewhat over the spectrum of basket securities, they share many common features. The approach taken here is to describe in some detail the characteristics and related benefits of the Standard & Poor's Depositary Receipt (SPDR), as the "granddaddy" of all currently trading baskets. Various differences from the SPDR are noted later in the discussion of other baskets.

CHARACTERISTICS AND BENEFITS OF THE STANDARD & POOR'S DEPOSITARY RECEIPT (SPDR)

Based upon the S&P 500 Index, the SPDR was introduced on January 22, 1993, with the initial issuance of 150,000 SPDRs, worth about $6 million. (As of May 31, 2001, SPDR assets amounted to about $28.7 billion.) The issuing entity is the SPDR Trust, a UIT registered under the 1940 Act. The sponsor is PDR Services LLC, solely owned by the American Stock Exchange LLC. The trustee is State Street Bank and Trust; and the distributor is ALPS Mutual Fund Services, Inc.

Portfolio Deposits and Redemptions

Not just anybody can deal directly with the SPDR Trust. SPDRs are issued by its trust in multiples of 50,000 SPDRs, each of which constitutes one "creation unit." Only those investors that have entered into a "participating agreement" with the SPDR Trust may transact in creation units. These participants who deal directly with the SPDR Trust are of necessity large investment organizations such as investment banking firms. Moreover, transactions with the SPDR Trust in creation units do not take place using cash. Rather, deposits to, and redemptions from, the SPDR Trust take place predominantly *in kind*, with the actual basket of stocks that make up the S&P 500 Index.

These deposit and redemption procedures are highly significant for two primary reasons. One benefit is a mechanism they provide to ensure that the price of the SPDR, as it trades during normal market hours on the American Stock Exchange (AMEX), remains close to the net asset value (NAV) of its underlying shares that are held in the SPDR Trust. This is so because of the potential for arbitrage profits to be made by participants should the price of SPDR deviate too far in either direction from its underlying value. Hence, the price of the exchange-traded SPDR (stock symbol SPY) is relatively **price efficient**. This is important to know because the typical investor, like most of us, cannot buy from or sell directly to the SPDR Trust.

A further benefit of the portfolio deposit and redemption features is **tax efficiency**. Because transactions directly with the SPDR Trust are conducted predominantly in kind, the Trust does not have to sell stocks inside the Trust to meet redemptions. And because the portfolio is an index fund, changes in portfolio holdings are made only when the index changes. So, unlike an open-end actively managed mutual fund, realized capital gains inside the SPDR Trust are minimal. As a result, a total of only about $0.09 in realized capital gains have been distributed to SPDR holders since inception in January 1993.

While not all basket securities are as efficient as the SPDR, price efficiency and tax efficiency are hallmark benefits of basket securities in general.

Exchange Trading

While creation unit transactions are the way the "Big Guys" deal with the SPDR Trust, trading of SPY on the AMEX is the route for the rest of us. A primary benefit is **liquidity**. In contrast to open-end mutual funds, with which transactions can be made only as of the market close each day at NAV, SPDRs trade during all normal market hours. SPDRs are priced at one-tenth of the level of the S&P 500 Index, so their price can be checked for efficiency. For example, if the S&P 500 Index is 1250, the price of SPY on the AMEX should be about $125. The Indicative Optimized Portfolio Value (IOPV), an approximated NAV of a SPDR, is calculated and disseminated about every 15 seconds during normal market hours, so if the forces of supply and demand (which determine the price for SPY in the short run) move the price too far from its NAV, the potential for arbitrage profits (discussed in the previous section) comes into the pricing relationship and can move the market price toward NAV. SPDRs are highly liquid and can be bought "in size." The minimum purchase (or sale) is one share. So liquidity is a major benefit that spans the spectrum of transaction size.

As additional benefits of their exchange-traded nature, SPDRs can be margined and can be sold short on a "downtick" in price, for those investors who are so inclined to engage in such activities. Margin is simply borrowing (typically from a broker) to buy more shares than the amount of cash available would otherwise support. Generally, regulations do not permit shorting a stock on a downtick because such transactions could potentially put undue price pressure on a single issue. However, the SEC permits shorting on downticks in SPY because SPDR represents all 500 stocks in the S&P 500 Index, whick makes it unlikely that such transactions could impact the price of the entire index.

COST EFFICIENCY

It was noted earlier that UITs do not have a board of directors. (They do have a sponsor and a trustee.) Because its portfolio redemption and deposit procedures deal only with the Big Guys, SPDR does not have a multitude of shareholder service representatives to answer inquiries from shareholders. Because it holds an index fund, the SPDR Trust does not have to pay for investment research. The result of all this is **cost efficiency**. The annual expense ratio for SPDR is currently 12 basis points, or 0.0012.

FRESHNESS

As portfolio managers, the authors often encounter situations in which a substantial portion of an individual's wealth is accounted for by an individual stock or just a few stocks. Typically, the stock has been owned for a long while (sometimes decades), and has a low cost basis for tax purposes. It is not unusual to find that the stock is not performing as well as the market as a whole. In such situations (and even if the stock is performing), greater diversification is called for in order to reduce the risk of the specific holding. Such diversification can be costly from the standpoint of taxes to be paid on realized capital gains. Even if a diversification program is spread over several years, the risk of the portfolio is higher by virtue of the concentration that remains.

By contrast, since the S&P 500 Index is updated from time to time, by deleting some stocks and adding others, it continues to represent the present economic structure over time. That is, its holdings are up-to-date, or fresh. Accordingly, not only does the SPDR provide the diversification of the entire index, its composition is likely to be as fresh decades from now as it is today. The risk/return profile of the portfolio described in the previous paragraph is strikingly different from a portfolio constructed around basket securities. More about such portfolio construction later, but for now it is useful to know that a benefit of baskets is **freshness**.

REDUCING BOMBSHELLS

Similar to freshness, the reduction of the occurrence of "bombshells" is another portfolio construction topic. Bombshell is the term we apply to a stock that blows up and declines severely in price. This can happen for any of a variety of disappointments related to sales, earnings, margins, growth rates, or even, in the case of a major software firm in 2000, the body language of the chief financial officer. Also, data confirms that bombshells are considerably more frequent since the end of 1999 than was previously the case. To the extent basket securities are used in portfolio construction, the exposure to potential damage from bombshells is reduced.

This chapter has explored a number of structural characteristics and their related benefits that are associated with basket securities. With a focus on the SPDR as a representative model, we have seen that the benefits of basket securities include instant diversification, price efficiency, tax efficiency, liquidity, cost efficiency, freshness, and the reduction of bombshells. Not bad for one single security!

3 A Survey of Major Participants

While Chapter 2 focused on the Standard & Poor's Depositary Receipt as a model for understanding the primary structural characteristics and benefits of basket securities generally, this chapter surveys the major participants presently involved in creating and operating various aspects of the basket securities market. These participants serve in one or more roles such as sponsor or investment manager/adviser. The reader is encouraged to use the data in Appendix A at the end of the book for further detail regarding the various baskets.

THE AMERICAN STOCK EXCHANGE (AMEX)

As described in Chapter 1 regarding the origins of basket securities, the AMEX has for a long time had an interest in the basket concept. An early attempt in 1989 with its Equity Index Participations failed due to the instruments being deemed by the U.S. Court of Appeals of the Seventh Circuit to be inherently futures contracts. Such contracts fall under the regulatory jurisdiction of the Commodity Futures Trading Commission, not the Securities and Exchange Commission; and so the product was withdrawn from the market.

Then in January 1992, just less than three months after the Index Trust SuperUnit was launched on the AMEX, the exchange introduced its own S&P 500 Index basket, the SPDR. This latter security has become the granddaddy of baskets (at least in terms of longevity) and was described in some detail in Chapter 2.

The AMEX is the sole owner of PDR Services LLC, which is the sponsor of the SPDR Trust, the UIT (unit investment trust) that holds the underlying stocks and issues exchange-traded securities that are based on the various Standard & Poor's indices. Through this sponsor, the AMEX has brought to market and currently trades a broad variety of basket securities (exchange-traded funds). It also trades many basket securities that are sponsored by other organizations, discussed below. The AMEX clearly dominates basket trading.

In addition to the SPDR, other **broad-based index** baskets trade on the AMEX. These include the Midcap SPDR, based on the S&P 400 Stock Index (trading symbol MDY) and Diamonds, based on the Dow Jones Industrial Average (DIA).

Also traded on AMEX is the Nasdaq 100 Tracking Stock (QQQ) that represents the largest 100 nonfinancial companies on Nasdaq. This security is issued by the Nasdaq 100 Trust; and the index is weighted about 75 percent in large technology companies. It has become dominant, in terms of asset growth and trading volume, among all the basket securities, even though it was not introduced until 1999.

According to *The Wall Street Journal,* as of July 2001 the QQQ had assets of about $23 billion and daily trading volume of about 67 million shares.

The AMEX (through PDR Services LLC) has also introduced and currently trades a number of SPDR **sector** baskets. Representation in economic sectors is available for basic industries, consumer services, consumer staples, cyclical/transportation, energy, financial, technology, and utilities.

Other sector baskets available on the AMEX are **streetTRACKS Sector Funds**. At present these include the Fortune e-50 Index Tracking Stock (FEF), streetTRACKS Morgan Stanley High Tech 35 Index Fund (MTK), and streetTRACKS Morgan Stanley Internet Index Fund (MII).

Virtually all other basket securities that have been introduced by *sponsors other than* the AMEX PDR Services LLC also trade on the AMEX. [Exceptions include the iShares S&P Global 100 that trades on the New York Stock Exchange and the OEX100 (S&P100) that trades on the Chicago Board Options Exchange.] Information on all these baskets can be found online at www.amex.com, under "Exchange Traded Funds."

BARCLAYS GLOBAL INVESTORS, N.A. (BGI)

BGI, together with its wholly owned subsidiary, Barclays Global Fund Advisors (BGFA), is the world's largest investment advisor of institutional investment assets, and the largest manager of indexed products. BGI's association with basket securities also dates from their inception. As described in the appendix regarding the SuperTrust, Wells Fargo Nikko Investment Advisors, which through a series of corporate changes has become a part of BGI, served for a time as the subadvisor (i.e., investment manager) for the Index Series of Capital Market Fund. Shares of the Index Series were the basis for the Index Trust SuperUnit which tracked the S&P 500 Stock Index and was the first successful basket security, predating the launch of AMEX's SPDR by nearly 3 months.

BGI has continued and, especially since about mid-2000, has accelerated its involvement with basket securities. When WEBS (World Equity Benchmark Shares) were introduced in April 1996, BGI became their investment manager through a previous entity (Wells Fargo Nikko Investment Advisors), and remains investment advisor today through BGFA. WEBS have been renamed as iShares MSCI Series and are offered through iShares, Inc., an index mutual fund consisting of separate series'. Each of these basket securities respectively tracks a particular index provided by Morgan Stanley Capital International Inc. (MSCI) for a geographic region, mostly individual foreign countries. They have traded on the AMEX since inception.

It seems only natural that BGI, the firm that now includes what was the investment operation of Wells Fargo, should be in the forefront of advancing the scope and role of basket securities/exchange-traded funds based upon indexing. *The Wall Street Journal,* in an April 9, 2001 review, "Index Funds: 25 Years in Pursuit of the Average," provided useful historical perspective regarding the development of this important market:

A Survey of Major Participants 15

"The earliest index portfolios grew out of the academic world's efficient-market theory, which postulates that stock prices are the product of all relevant information and investor expectations. That view suggests investors cannot reliably predict which stocks will outperform in the future. As far back as the 1960s, pioneering studies of the actual performance of pension funds and mutual funds also showed many managers trailing market benchmarks, such as the S&P 500.

From those **foundations** grew 'a very controversial, antiestablishment kind of idea' recalls William Fouse, who helped develop early index portfolios at Wells Fargo Bank in San Francisco. The idea: that investors could beat most 'active' stock pickers simply by owning a portfolio that holds the stocks contained in an index like the S&P 500.

In 1971, Wells Fargo created what many view as the first indexed portfolio…"

So, what seems to be in store for the future of index investing? The *WSJ* article goes on to say:

"…many indexing advocates believe the mechanical investing approach of indexing will claim a larger market share over time. How big? At Barclays Global Investors in San Francisco, which traces its corporate lineage to an early indexing unit of Wells Fargo, global chief executive Patricia Dunn thinks 50% of assets is not unreasonable. At that point, she says, investors in the aggregate would be 'agnostic' about whether 'passive' index investing or 'active' stockpicking is the winning strategy.

Of course, it isn't necessarily an either-or choice. Even indexing-uber-advocate Vanguard, which also runs stockpicking funds, sees virtue in an active-passive mix. 'Is it smart to track the market and try to beat it?' asks Vanguard Web ads; yes, they say, recommending active and passive funds 'working in tandem'. (Emphasis added; see Chapter 4 for the authors' suggestions regarding active-passive applications using basket securities.)

Ms. Dunn's company is currently at the center of one of the most interesting and debated developments in indexing — the proliferation of exchange-traded index funds. These funds can be bought and sold, most of them on the American Stock Exchange, throughout the trading day like individual stocks. Barclays last year introduced more than 40 exchange-traded iShares, many focusing on a single industry or country."

In addition to the iShares MSCI Series (that initially came to market in 1996 as WEBS — World Equity Benchmark Shares), BGI has entered into licensing agreements with additional "Index Providers" for the purpose of offering many additional basket securities based on a variety of indices.

Through the iShares Trust, a registered mutual fund with about 35 separate investment portfolios called Funds, BGI has introduced basket securities as follows: (See Appendix A for information on each of the securities.)

- iShares S&P Series, from indices provided by Standard & Poor's (a division of The McGraw-Hill Companies, Inc.)
- iShares Dow Jones Series, from indices provided by Dow Jones & Company
- iShares Russell Series, from indices provided by Frank Russell Company
- iShares Cohen & Steers Series, from indices provided by Cohen & Steers Capital Management, Inc.
- iShares Nasdaq Series, from indices provided by Nasdaq Stock Market, Inc.

THE NEW YORK STOCK EXCHANGE (NYSE)

To date, the NYSE, the world's largest stock exchange, has had limited involvement in the world of basket securities. In April 1996, the same month during which WEBS were introduced on the American Stock Exchange, **Country Baskets** began trading on the NYSE. Similar to WEBS, nine Country Basket securities were introduced to track indices in Australia, France, Germany, Hong Kong, Italy, Japan, South Africa, United Kingdom, and United States. The underlying indices were the *Financial Times*/Standard & Poor's Actuaries World Indices for each of the countries, respectively. The investment adviser/manager was Deutsche Morgan Grenfel, a wholly owned indirect subsidiary of Deutsche Bank AG, a major German banking institution. After a few months of trading, the Country Basket securities were discontinued, while WEBS went on to success.

The NYSE's only currently trading (and listed) basket security is the iShares S&P Global 100. Introduced in December 2000, it is based on the S&P Global 100 Index that was developed jointly by Standard & Poor's and the NYSE to identify "...100 leading companies whose performance is tied more to global economic, competitive, and industry trends than headquarter locations." The investment manager is BGI's Barclays Global Fund Advisors.

However, a *Wall Street Journal* feature on July 12, 2001, reported the imminent inception of trading on the NYSE of the Nasdaq 100 Tracking Stock (QQQ), as well as SPDR (SPY) and Diamonds (DIA), that track the Dow Jones Industrial Average. Such expansion of NYSE trading is seen as a further endorsement of the importance and usefulness of basket securities.

MERRILL LYNCH, PIERCE, FENNER & SMITH INCORPORATED (MERRILL LYNCH)

Merrill Lynch has launched a series of securities called HOLDRS, for **HOL**ding Company **D**epositary **R**eceipt**S** that trade on the American Stock Exchange. These securities are not based on indices, but rather are composed of a group of stocks (initially 20 per HOLDRS) that are selected by Merrill Lynch from within various industries or sectors. The composition of each group of stocks is fixed, and will not change except when a "reconstitution event" (such as a merger or spinoff) occurs. Presently numbering about 16 (see Appendix A for details), examples include Biotech, Broadband, Oil Services, Regional Bank, Semiconductor, and Wireless.

A Survey of Major Participants 17

HOLDRS differ from other basket securities discussed previously in several significant ways, such as:

- They are not based on indices, as noted above.
- They are not issued by investment companies registered under the Investment Company Act of 1940. Rather, securities of each respective HOLDRS are issued pursuant to a Depositary Trust Agreement under which The Bank of New York is trustee, and Merrill Lynch serves as the initial depositor of securities into each trust and as the leader of the selling group that offers securities issued by each trust to the public.
- HOLDRS can be purchased or sold only in round lots of 100 shares, or multiple round lots. In contrast, other basket securities reviewed (e.g., SPDRs) can be purchased or sold in odd lots as small as one share.
- An owner of HOLDRS may cancel the security by having it delivered to the trustee, in exchange for the underlying stocks. An investor may also create a HOLDRS security by delivering to the trustee the underlying basket of stocks. These procedures are different from the creation unit approach used by the other baskets securities considered. The latter typically require deposits and redemptions to be made in one or more creation units (e.g., each a minimum lot of 50,000 securities) by a participating party that has entered into an agreement with the trust.
- If an investor cancels a HOLDRS as above, the proportional shares received will have cost bases equal to the market values of the respective shares as of the date the HOLDRS was purchased.

In the next chapter we explore ways that basket securities can be useful in the investment management process.

Section II

Investment Strategies and Examples

4 Asset Management Strategies

THE INVESTMENT PROCESS GENERALLY

The authors assert that, for any investor, the essential steps in the investment process are the same. Moreover, the process should be undertaken explicitly, although (unfortunately) too often it is discernible only implicitly. The investment process begins with consideration of purpose, or objective, taking into account such things as income and liquidity needs, growth aspirations, time horizon, and risk tolerance. These considerations translate initially into a *strategic* asset allocation, an appropriate (longer-term) mix of the various asset categories such as cash, bonds, stocks, and perhaps real estate, although many times the topic is limited to financial assets.

The discussion of investment management strategies in this book presumes that the reader has **already addressed and reached an appropriate asset allocation decision** regarding the various financial asset categories. Accordingly, this chapter describes an alternative approach to the traditional means of achieving effective *diversification* in the construction of the equity portion of the portfolio.

USES OF BASKET SECURITIES IN EQUITY PORTFOLIOS

Trading the Market

Basket securities such as Standard and Poor's Depositary Receipts (the SPDR) that represent the S&P 500 Stock Index can be bought and sold on the American Stock Exchange (the AMEX) during all normal market hours. For those who may wish to trade the market, the SPDR, as well as the MidCap SPDR (the S&P 400 Stock Index) and the Nasdaq 100 Tracking Stock (the Nasdaq 100 Stock Index, composed of the 100 largest nonfinancial stocks on the Nasdaq), provide single-stock access to indices that trade efficiently.

As an Equity Portfolio

The three basket securities mentioned above, being based on indices, can be used **as** an entire equity portfolio. For example, the SPDR offers the risk and return characteristics of the entire S&P 500 Index. Blended together, as described below, the three baskets can provide the diversification of nearly 1,000 different companies, though care should be exercised in how one blends them.

Within an Equity Portfolio

Basket securities can be used in many ways **within** an equity portfolio, for example:

- To create a diversified core, using some combination of index baskets, e.g. the SPDR (exchange trading symbol SPY), the MidCap SPDR (MDY), and the Nasdaq 100 Tracking Stock (QQQ).
- To augment sector exposure, using Select Sector SPDR Funds, or iShares Sector Funds (more on this in Chapter 5). The SPDR offerings include sector exposures to basic industries (XLB), consumer staples (XLP), consumer services (XLV), energy (XLE), financial (XLF), industrial (XLI), technology (XLK), transportation (XLY), and utilities (XLU). Sector iShares offer an even broader array of exposures, based on Dow Jones U.S. indices. A good place to look (in addition to Appendix A of this book) for more information and listings of holdings of the sector baskets is the AMEX website, at www.amex.com (click on Exchange Traded Funds, the Sector Indexes). Technology exposure can also be effectively managed by using the QQQ.

The Thousand Stock Portfolio

The "Thousand Stock Portfolio" which, as we will see, holds only some *50 or so securities*, is offered here as a further example of an application of basket securities that the authors are using. Its primary attributes include:

- Effective diversification — a key feature of any well-constructed equity portfolio over time
- A recognition that at any point in time portfolio managers (or investors generally, although what follows is described in the context of professional portfolio managers) seldom, if ever, have the same degree of confidence in, or enthusiasm for, each and every stock in the portfolio
- A "semipassive" diversified core, combined with "active" equity selections.

The following sections explore each of these features in some detail.

Diversification

The traditional approach to equity portfolio diversification is to invest in stocks within the various economic industries or sectors, and perhaps with various size (capitalization) and style (e.g. growth, value) considerations. This approach, of course, continues to work effectively. A traditional portfolio comprised of 40 to 50 stocks or so, constructed around these various criteria, can provide good diversification, as long as extreme exposures are not taken relative to some market index or performance benchmark.

As an alternative with some possible advantages (in the authors' opinion), the Thousand Stock Portfolio approaches diversification in a somewhat different manner. Basket securities are used to create a **diversified core**. For example, if one blends

Asset Management Strategies

the SPDR (SPY), the MidCap SPDR (MDY), and the Nasdaq 100 Tracking Stock (QQQ), nearly 1,000 different companies are represented, as follows:

- SPY represents 500 stocks, the S&P 500 Index.
- MDY represents 400 stocks, the S&P 400 Index.
- QQQ represents 100 stocks, the Nasdaq 100 Index.

However, because there are some stocks that are included in both the S&P 500 Index and the Nasdaq 100 Index, the combination of these three baskets provides exposure to somewhat fewer than 1,000 *different* companies, currently 937 with a range on the order of perhaps 925–950.

The three baskets described above can be blended in various ways, depending on considerations such as the aggressiveness of the portfolio and the perceived opportunity presented by each basket. The **portion** of the equity portfolio that is invested in the diversified core can also vary, depending on the aggressiveness of the portfolio and the number of active stock ideas the portfolio manager may have from time to time. As described below, variations made to the diversified core, make it semiactive, even though it is constructed with index-related securities.

The portfolio manager or investor needs a way to measure and manage the exposures to various economic sectors that are contributed by the particular mix of the baskets being used. This is so because, when individual stocks are included in the portfolio, the *overall* exposures to economic sectors need to be measured and controlled, as one important aspect of diversification.

The following spreadsheets provide a means of measuring and controlling exposures to economic sectors. The first spreadsheet (Figure 4.1) shows, as of December 31, 2000, the economic sector contributions of a diversified core comprising **60 percent** of the equity portfolio, with the three baskets blended as:

1. 50 percent SPY, (SPDR, S&P 500)
2. 35 percent MDY, (MidCap SPDR, S&P 400)
3. 15 percent QQQ, (Nasdaq 100)

The spreadsheet shown in Figure 4.2 is similar, except that the diversified core comprises 75 percent of the equity portfolio. In each case, the remaining portion of the portfolio (40 percent for the first, 25 percent for the second) is available for the portfolio manager's "best" active ideas. An advantage of this approach is that the portfolio manager can be less concerned about (but cannot ignore) the impact on overall equity portfolio diversification that the individual stock portion makes. That is, the primary focus of the portfolio manager can be on issues that are believed to have the potential to enhance investment results, as contrasted with individual stock selections that are based largely upon diversification needs.

To illustrate, suppose the diversified core is as shown in the second spreadsheet above, with 75 percent in the three baskets. Suppose further that the portfolio manager believes that a number of technology stocks are likely to rebound in the year or so ahead. As a result, the portfolio manager, taking into consideration the risk tolerance and other characteristics of the client, concludes that an overweighting

"Thousand Stock Portfolio"

Account Number: _____
Account Name: SAMPLE; 60% BASKETS
Date: _____

SECTORS	% of Portfolio: 30.00% %SPY	Contribution	% of Portfolio: 21.00% %MDY	Contribution	% of Portfolio: 9.00% %QQQ	Contribution	TOTAL FROM BASKETS	Sum of % Portfolio: 60.00%
Basic Materials	2.60%	0.78%	4.10%	0.86%	0.30%	0.03%	1.67%	
Capital Goods	8.90%	2.67%	7.20%	1.51%	2.01%	0.18%	4.36%	
Communication Svcs.	6.00%	1.80%	1.50%	0.32%	6.41%	0.58%	2.69%	
Consumer Cyclicals	8.60%	2.58%	14.70%	3.09%	4.71%	0.42%	6.09%	
Consumer Staples/Services	13.10%	3.93%	7.70%	1.62%	4.85%	0.44%	5.98%	
Energy	7.10%	2.13%	7.20%	1.51%		0.00%	3.64%	
Financial Svcs.	17.40%	5.22%	16.40%	3.44%		0.00%	8.66%	
Health Care	12.90%	3.87%	12.30%	2.58%	10.50%	0.95%	7.40%	
Technology	18.60%	5.58%	19.00%	3.99%	71.12%	6.40%	15.97%	
Transportation	0.70%	0.21%	1.90%	0.40%	0.10%	0.01%	0.62%	
Utilities-Gas & Elec.	4.10%	1.23%	8.00%	1.68%		0.00%	2.91%	
TOTAL	100.00%		100.00%		100.00%		60.00%	

SECTORS	% of Individual Stocks	Total for Portfolio
Basic Materials		1.67%
Capital Goods		4.36%
Communication Svcs.		2.69%
Consumer Cyclicals		6.09%
Consumer Staples/Services		5.98%
Energy		3.64%
Financial Svcs.		8.66%
Health Care		7.40%
Technology		15.97%
Transportation		0.62%
Utilities-Gas & Elec.		2.91%
TOTAL	0.00%	60.00%

FIGURE 4.1 Diversified Core as 60 Percent of the Equity Portfolio.

Asset Management Strategies

"Thousand Stock Portfolio"

Account Number: _____
Account Name: SAMPLE; 75% BASKETS
Date: _____

SECTORS	% of Portfolio: 37.50% %SPY	Contribution	% of Portfolio: 26.25% %MDY	Contribution	% of Portfolio: 11.25% %QQQ	Contribution	TOTAL FROM BASKETS	Sum of % Portfolio: 75.00%
Basic Materials	2.60%	0.98%	4.10%	1.08%	0.30%	0.03%	2.09%	
Capital Goods	8.90%	3.34%	7.20%	1.89%	2.01%	0.23%	5.45%	
Communication Svcs.	6.00%	2.25%	1.50%	0.39%	6.41%	0.72%	3.36%	
Consumer Cyclicals	8.60%	3.23%	14.70%	3.86%	4.71%	0.53%	7.61%	
Consumer Staples/Services	13.10%	4.91%	7.70%	2.02%	4.85%	0.55%	7.48%	
Energy	7.10%	2.66%	7.20%	1.89%		0.00%	4.55%	
Financial Svcs.	17.40%	6.53%	16.40%	4.31%		0.00%	10.83%	
Health Care	12.90%	4.84%	12.30%	3.23%	10.50%	1.18%	9.25%	
Technology	18.60%	6.98%	19.00%	4.99%	71.12%	8.00%	19.96%	
Transportation	0.70%	0.26%	1.90%	0.50%	0.10%	0.01%	0.77%	
Utilities-Gas & Elec.	4.10%	1.54%	8.00%	2.10%		0.00%	3.64%	
TOTAL	100.00%		100.00%		100.00%		75.00%	

SECTORS	% of Individual Stocks	Total for Portfolio
Basic Materials		2.09%
Capital Goods		5.45%
Communication Svcs.		3.36%
Consumer Cyclicals		7.61%
Consumer Staples/Services		7.48%
Energy		4.55%
Financial Svcs.		10.83%
Health Care		9.25%
Technology		19.96%
Transportation		0.77%
Utilities-Gas & Elec.		3.64%
TOTAL	0.00%	75.00%

FIGURE 4.2 Diversified Core as 75 Percent of the Equity Portfolio.

"Thousand Stock Portfolio"

Account Number: ___
Account Name: SAMPLE; 75% BASKETS PLUS STOCKS
Date: ___

SECTORS	% of Portfolio: %SPY	37.50% Contribution	% of Portfolio: %MDY	26.25% Contribution	% of Portfolio: %QQQ	11.25% Contribution	TOTAL FROM BASKETS	Sum of % Portfolio: 75.00%
Basic Materials	2.60%	0.98%	4.10%	1.08%	0.30%	0.03%	2.09%	
Capital Goods	8.90%	3.34%	7.20%	1.89%	2.01%	0.23%	5.45%	
Communication Svcs.	6.00%	2.25%	1.50%	0.39%	6.41%	0.72%	3.36%	
Consumer Cyclicals	8.60%	3.23%	14.70%	3.86%	4.71%	0.53%	7.61%	
Consumer Staples/Services	13.10%	4.91%	7.70%	2.02%	4.85%	0.55%	7.48%	
Energy	7.10%	2.66%	7.20%	1.89%		0.00%	4.55%	
Financial Svcs.	17.40%	6.53%	16.40%	4.31%		0.00%	10.83%	
Health Care	12.90%	4.84%	12.30%	3.23%	10.50%	1.18%	9.25%	
Technology	18.60%	6.98%	19.00%	4.99%	71.12%	8.00%	19.96%	
Transportation	0.70%	0.26%	1.90%	0.50%	0.10%	0.01%	0.77%	
Utilities-Gas & Elec.	4.10%	1.54%	8.00%	2.10%		0.00%	3.64%	
TOTAL	100.00%		100.00%		100.00%		75.00%	

SECTORS	% of Individual Stocks	Total for Portfolio		
Basic Materials	2.20%	4.29%	Alcoa	
Capital Goods	4.60%	10.05%	General Electric, Tyco	
Communication Svcs.	2.20%	5.56%	WorldCom	
Consumer Cyclicals	2.20%	9.81%	Walmart	
Consumer Staples/Services	2.20%	9.68%	Pepsi	
Energy	2.20%	6.75%	ExxonMobil	
Financial Svcs.	2.20%	13.03%	Citigroup	
Health Care	2.20%	11.45%	Johnson & Johnson	
Technology	5.00%	24.96%	IBM, EMC, CiscoSystems	
Transportation		0.77%		
Utilities-Gas & Elec.		3.64%		
TOTAL	25.00%	100.00%		

FIGURE 4.3 Diversified Core as 75 Percent of the Equity Portfolio.

Asset Management Strategies 27

of the technology sector is desirable, and that such weighting could be up to 25 percent of the equity portfolio, compared to about a 19 percent weighting of the technology sector in the S&P 500 Index (the presumed market proxy/benchmark being used in this situation).

The portfolio manager selects several technology stocks to hold in the portfolio, together with other stocks selected for their perceived potential. (Sector basket securities can also be used to assist in managing overall equity portfolio diversification; see Appendix A). The *total* equity portfolio might look similar to the one illustrated in Figure 4.3, including the individual stocks selected.

As time passes and conditions change, the portfolio manager may make changes among individual stock holdings, may have more or fewer good ideas at any point in time, and can reflect these ideas in the portfolio. If the portfolio manager has more active (i.e., good) ideas, the diversified core can be drawn upon as a source of funds for such ideas, thus reducing it somewhat. If, on the other hand, the portfolio manager has fewer active ideas to put into the portfolio at some point in time, proceeds from the sale of individual stocks can be added to the diversified core. In this way the diversified core can be used as a sort of inventory fund for providing resources to, or absorbing proceeds from, the active component as the portfolio manager's views change over time.

The significance of this approach is that the portfolio manager does not have to sell a stock in order to buy a new one; the diversified core can supply the needed resources. Likewise, the portfolio manager does not have to find a new idea in order to sell a stock; the diversified core can absorb the proceeds.

Hence, the diversified core is in reality semiactive. Although it is composed of index-related securities, both the *blend* of the various basket securities, and the *portion* of the equity portfolio they constitute at a particular time, are based on deliberate decisions, as described in the previous paragraphs. These issues are also addressed in the next chapter.

This combination of a **semiactive diversified core,** plus active equity selections (i.e., the Thousand Stock Portfolio) gives the portfolio manager sufficient flexibility to manage the portfolio across a spectrum of aggressiveness, relative to the client's parameters and the portfolio manager's degree of confidence/enthusiasm for specific individual stocks or sectors at any point in time. Simultaneously, effective diversification can be maintained.

5 An Example: Thousand Stock Portfolio

Basket securities, representing passive index funds, can be used as a total equity portfolio. Alternatively, in those situations where the investor has a basis for engaging in active management, basket securities can be used to create a diversified core. Active investment selections can be combined with the diversified core to create a blend of passive/active management. Active investment decisions can be made and implemented with individual stocks or other securities such as basket securities that represent various sectors of the economy.

It is important to note that this book is not about active management. For readers who are not investment professionals, active management using individual stocks is not recommended. The specific risk of individual stock selections is high and, if the reader lacks experience, it is best left to the professionals (who, by the way, often also do not have an easy time selecting winning stocks). The next title in the Power Investing series, *Power Investing with Stocks*, will deal with this topic of individual stock selection and discuss several modern tools for selection, portfolio construction, and risk control.

Professional investment managers follow innumerable approaches to stock selection. For example, the authors in their everyday work use a combination of quantitative, fundamental, and timeliness factors to select stocks for clients. These various factors are derived from a host of sources, including top-down strategic thinking, industry and company research from both buy-side and sell-side research analysts, and several proprietary research services that provide the opportunity to develop insights regarding a given company's past and likely future economic performance, the valuation of the stock, and the level of expectations being priced into the stock by the market. Despite all the tools and resources available, successful stock selection remains as much an art as a science. We would be doing a disservice to the nonprofessional investor to recommend in this book that he or she undertake active stock picking to augment the diversified core of index basket securities.

However, for the investment professional who has a basis for active management, and for the nonprofessional who wishes to go beyond the diversified core in some fashion, the material that follows describes an approach for the use of select sector SPDR baskets that are combined with the diversified core. The professional could use individual stocks together with sector baskets to make up the active portion of the portfolio. We suggest that the nonprofessional limit active exposure to the sector baskets.

To restate, the Thousand Stock Portfolio described in Chapter 4 is composed of a semipassive diversified core, combined with active equity selections (individual stocks or sector baskets). This chapter introduces a simple and highly useful application of this strategy using only sector baskets as the active portion. Before the availability of these basket securities, successful variations of this approach were employed for many years using sector mutual funds.

Once again, assume that the overall portfolio asset allocation decision has already been determined, that is, the percentage allocation to equities, fixed-income (taxable or tax-free), and cash/equivalents has been made. This example targets the equity portion of the asset mix.

This application of a Thousand Stock Portfolio strategy will be referred to as the "Basket Case Portfolio," or BCP. As described in the last chapter, the passive portion of the Basket Case Portfolio uses a diversified core comprising the three index baskets: the S&P 500 SPDR (SPY), the MidCap SPDR (MDY), and the Nasdaq 100 Tracking Stock (QQQ).

Rather than using individual securities, the active segment of the BCP portfolio is created from the select sector SPDR funds offering sector exposures to basic industries (XLB), consumer staples (XLP), consumer services (XLV), energy (XLE), financial (XLF), industrial (XLI), technology (XLK), transportation (XLY), and utilities (XLU).

The total investable universe consists of 12 exchange-traded securities distributed between two accounts, or portions of the equity portfolio: the diversified core account and the sector SPDR account. These two accounts constitute the passive and active portions of the portfolio respectively. The portfolio benchmark is the Standard & Poor's 500 Index. Table 5.1 lists all of the SPDR basket securities currently available. While this example is based on the SPDR basket securities, the reader should be aware that other baskets, both of the index type and the sector type are available, primarily as iShares. The iShares offerings may be used as alternatives to or complements for SPDR securities in constructing a BCP.

TABLE 5.1
The Basket Case Universe of Funds

Passive Allocation — Diversified Core Funds

Fund Name	Ticker	Representative Index
SPDR Trust Series 1	SPY	S&P 500 Index
MidCap SPDR Trust Series 1	MDY	S&P 400 Index
Nasdaq-100 Trust Series 1	QQQ	Nasdaq 100 Index

Active Allocation — S&P Sector SPDRs

Fund Name	Ticker	Representative Index
Basic Industries Select SPDR	XLB	S&P Basic Industries
Consumer Services Select SPDR	XLV	S&P Consumer Services
Consumer Staples Select SPDR	XLP	S&P Consumer Staples
Cyclical/Transportation Select SPDR	XLY	S&P Cyclical/Transportation
Energy Select SPDR	XLE	S&P Energy
Financial Select SPDR	XLF	S&P Financial
Industrial Select SPDR	XLI	S&P Industrial
Technology Select SPDR	XLK	S&P Technology
Utilities Select SPDR	XLU	S&P Utilities

An Example: Thousand Stock Portfolio

Tables 5.2 and 5.3 list all the companies included in the baskets that constitute the BCP, sorted by index percentage allocation, as of March 31, 2001.

Diversification is effectively dealt with in the Diversified Core Account as 937 companies are combined in the three basket securities — SPY, MDY, and QQQ. The complete listing is supplied in order to convey a sense of the scope of the portfolio diversification.

In the following tables, the S&P 500 securities are grouped by inclusion in the nine select sector SPDR funds.

TABLE 5.2
Companies in the Diversified Core Account

Ticker	Description	% Allocation	Sector
	S&P 500 SPDR (SPY)		
GE	General Elec Co.	4.0%	Capital Goods
MSFT	Microsoft Corp.	2.8%	Technology
XOM	Exxon Mobil Corp.	2.7%	Energy
PFE	Pfizer Inc.	2.5%	Health Care
C	Citigroup Inc.	2.2%	Financials
WMT	Wal Mart Stores Inc.	2.2%	Consumer Cyclicals
AIG	American Intl. Group Inc.	1.8%	Financials
INTC	Intel Corp.	1.7%	Technology
MRK	Merck & Co. Inc.	1.7%	Health Care
AOL	AOL Time Warner Inc.	1.7%	Consumer Staples
IBM	International Business Machs	1.6%	Technology
SBC	SBC Communications Inc.	1.5%	Communication Services
VZ	Verizon Communications	1.3%	Communication Services
JNJ	Johnson & Johnson	1.2%	Health Care
RD	Royal Dutch Pete Co.	1.1%	Energy
BMY	Bristol Myers Squibb Co.	1.1%	Health Care
CSCO	Cisco Sys Inc.	1.1%	Technology
KO	Coca Cola Co.	1.1%	Consumer Staples
MO	Philip Morris Cos Inc.	1.0%	Consumer Staples
HD	Home Depot Inc.	1.0%	Consumer Cyclicals
BAC	Bank Amer Corp.	0.9%	Financials
LLY	Lilly Eli & Co.	0.8%	Health Care
JPM	J P Morgan Chase & Co.	0.8%	Financials
WFC	Wells Fargo & Co. New	0.8%	Financials
ORCL	Oracle Corp.	0.8%	Technology
PG	Procter & Gamble Co.	0.8%	Consumer Staples
T	AT & T Corp.	0.8%	Communication Services
FNM	Federal Natl Mtg Assn	0.8%	Financials
AHP	American Home Products Corp.	0.7%	Health Care
VIA/B	Viacom Inc.	0.7%	Consumer Staples
BLS	Bellsouth Corp.	0.7%	Communication Services

(*continued*)

TABLE 5.2 (CONTINUED)
Companies in the Diversified Core Account

Ticker	Description	% Allocation	Sector
TYC	Tyco Intl. Ltd New	0.7%	Capital Goods
ABT	Abbott Labs	0.7%	Health Care
DELL	Dell Computer Corp.	0.6%	Technology
PHA	Pharmacia Corp.	0.6%	Health Care
EMC	E M C Corp. Mass	0.6%	Technology
PEP	Pepsico Inc.	0.6%	Consumer Staples
AMGN	Amgen Inc.	0.6%	Health Care
HWP	Hewlett Packard Co.	0.6%	Technology
MWD	Morgan Stanley Dean Witter & Co.	0.6%	Financials
DIS	Disney Walt Co.	0.6%	Consumer Staples
Q	Qwest Communications Intl. Inc.	0.6%	Communication Services
CHV	Chevron Corp.	0.5%	Energy
MDT	Medtronic Inc.	0.5%	Health Care
AXP	American Express Co.	0.5%	Financials
TXN	Texas Instrs Inc.	0.5%	Technology
WCOM	Worldcom Inc.	0.5%	Communication Services
SGP	Schering Plough Corp.	0.5%	Health Care
F	Ford Mtr Co. Del	0.5%	Consumer Cyclicals
SUNW	Sun Microsystems Inc.	0.5%	Technology
BA	Boeing Co.	0.5%	Capital Goods
FRE	Federal Home Ln Mtg Corp.	0.4%	Financials
MER	Merrill Lynch & Co. Inc.	0.4%	Financials
NT	Nortel Networks Corp.	0.4%	Technology
USB	Us Bancorp Del	0.4%	Financials
ENE	Enron Corp.	0.4%	Utilities
QCOM	Qualcomm Inc.	0.4%	Technology
DD	Du Pont E I De Nemours & Co.	0.4%	Basic Materials
ONE	Bank One Corp.	0.4%	Financials
BUD	Anheuser Busch Cos Inc.	0.4%	Consumer Staples
WAG	Walgreen Co.	0.4%	Consumer Staples
FBF	Fleetboston Finl Corp.	0.4%	Financials
MMM	Minnesota Mng & Mfg Co.	0.4%	Capital Goods
CMCSK	Comcast Corp.	0.4%	Consumer Staples
TX	Texaco Inc.	0.4%	Energy
KMB	Kimberly Clark Corp.	0.4%	Consumer Staples
BK	Bank New York Inc.	0.4%	Financials
AMAT	Applied Materials Inc.	0.3%	Technology
MCD	McDonalds Corp.	0.3%	Consumer Staples
ADP	Automatic Data Processing Inc.	0.3%	Technology
UTX	United Technologies Corp.	0.3%	Capital Goods
LU	Lucent Technologies Inc.	0.3%	Technology
G	Gillette Co.	0.3%	Consumer Staples
SLB	Schlumberger Ltd	0.3%	Energy

TABLE 5.2 (CONTINUED)
Companies in the Diversified Core Account

Ticker	Description	% Allocation	Sector
DUK	Duke Energy Co.	0.3%	Utilities
EPG	El Paso Corp.	0.3%	Utilities
TGT	Target Corp.	0.3%	Consumer Cyclicals
HON	Honeywell Intl. Inc.	0.3%	Capital Goods
FTU	First Un Corp.	0.3%	Financials
CCU	Clear Channel Communications	0.3%	Consumer Staples
WM	Washington Mut Inc.	0.3%	Financials
CL	Colgate Palmolive Co.	0.3%	Consumer Staples
MOT	Motorola Inc.	0.3%	Technology
AA	Alcoa Inc.	0.3%	Basic Materials
ALL	Allstate Corp.	0.3%	Financials
CPQ	Compaq Computer Corp.	0.3%	Technology
UN	Unilever N V	0.3%	Consumer Staples
GM	General Mtrs Corp.	0.3%	Consumer Cyclicals
KRB	Mbna Corp.	0.3%	Financials
DOW	Dow Chem Co.	0.3%	Basic Materials
HI	Household Intl. Inc.	0.3%	Financials
SWY	Safeway Inc.	0.3%	Consumer Staples
BAX	Baxter Intl. Inc.	0.3%	Health Care
CAH	Cardinal Health Inc.	0.3%	Consumer Staples
EMR	Emerson Elec Co.	0.3%	Capital Goods
AES	AES Corp.	0.3%	Utilities
MMC	Marsh & McLennan Cos Inc.	0.3%	Financials
EDS	Electronic Data Sys Corp. New	0.3%	Technology
FITB	Fifth Third Bancorp	0.2%	Financials
MU	Micron Technology Inc.	0.2%	Technology
JDSU	JDS Uniphase Corp.	0.2%	Technology
SO	Southern Co.	0.2%	Utilities
FDC	First Data Corp.	0.2%	Technology
MET	MetLife Inc.	0.2%	Financials
CVS	CVS Corp.	0.2%	Consumer Staples
HCA	HCA Healthcare Co.	0.2%	Health Care
LOW	Lowes Cos Inc.	0.2%	Consumer Cyclicals
SCH	Schwab Charles Corp.	0.2%	Financials
EXC	Exelon Corp.	0.2%	Utilities
KR	Kroger Co.	0.2%	Consumer Staples
KSS	Kohls Corp.	0.2%	Consumer Cyclicals
WMB	Williams Cos Inc.	0.2%	Utilities
GPS	Gap Inc.	0.2%	Consumer Cyclicals
MEL	Mellon Finl Corp.	0.2%	Financials
AGC	American Gen Corp.	0.2%	Financials
PNC	Pnc Finl Svcs Group Inc.	0.2%	Financials

(*continued*)

TABLE 5.2 (CONTINUED)
Companies in the Diversified Core Account

Ticker	Description	% Allocation	Sector
STI	Suntrust Bks Inc.	0.2%	Financials
FON	Sprint Corp.	0.2%	Communication Services
GLW	Corning Inc.	0.2%	Technology
VRTS	Veritas Software Corp.	0.2%	Technology
UNH	UnitedHealth Group Inc.	0.2%	Health Care
SLE	Sara Lee Corp.	0.2%	Consumer Staples
SYY	Sysco Corp.	0.2%	Consumer Staples
PCS	Sprint Corp.	0.2%	Communication Services
COST	Costco Whsl Corp. New	0.2%	Consumer Cyclicals
COC/B	Conoco Inc.	0.2%	Energy
IP	International Paper Co.	0.2%	Basic Materials
ITW	Illinois Tool Wks Inc.	0.2%	Capital Goods
TLAB	Tellabs Inc.	0.2%	Technology
DYN	Dynegy Inc. New	0.2%	Utilities
AT	Alltel Corp.	0.2%	Communication Services
CI	Cigna Corp.	0.2%	Health Care
HAL	Halliburton Co.	0.2%	Energy
CCL	Carnival Corp.	0.2%	Consumer Cyclicals
NCC	National City Corp.	0.2%	Financials
CA	Computer Assoc Intl. Inc.	0.2%	Technology
LEH	Lehman Brothers Hldgs Inc.	0.2%	Financials
APC	Anadarko Pete Corp.	0.2%	Energy
LMT	Lockheed Martin Corp.	0.2%	Capital Goods
GCI	Gannett Inc.	0.2%	Consumer Cyclicals
CPN	Calpine Corp.	0.2%	Utilities
CAT	Caterpillar Inc.	0.2%	Capital Goods
WMI	Waste Mgmt Inc. Del	0.2%	Capital Goods
STT	State Street Corporation	0.2%	Financials
D	Dominion Res Inc. Va New	0.2%	Utilities
AEP	American Elec Pwr Inc.	0.1%	Utilities
AFL	AFLAC Inc.	0.1%	Financials
OMC	Omnicom Group	0.1%	Consumer Cyclicals
P	Phillips Pete Co.	0.1%	Energy
BBT	BB&T Corp.	0.1%	Financials
THC	Tenet Healthcare Corp.	0.1%	Health Care
GDT	Guidant Corp.	0.1%	Health Care
PAYX	Paychex Inc.	0.1%	Technology
PVN	Providian Finl Corp.	0.1%	Financials
A	Agilent Technologies Inc.	0.1%	Technology
HNZ	Heinz H J Co.	0.1%	Consumer Staples
UNP	Union Pac Corp.	0.1%	Transportation
NTRS	Northern Trust Corp.	0.1%	Financials
REI	Reliant Energy Inc.	0.1%	Utilities

An Example: Thousand Stock Portfolio

TABLE 5.2 (CONTINUED)
Companies in the Diversified Core Account

Ticker	Description	% Allocation	Sector
LUV	Southwest Airls Co.	0.1%	Transportation
HIG	Hartford Financial Svcs Grp	0.1%	Financials
RIG	Transocean Sedco Forex Inc.	0.1%	Energy
ABS	Albertsons Inc.	0.1%	Consumer Staples
ADI	Analog Devices Inc.	0.1%	Technology
LLTC	Linear Technology Corp.	0.1%	Technology
CPB	Campbell Soup Co.	0.1%	Consumer Staples
GD	General Dynamics Corp.	0.1%	Capital Goods
CB	Chubb Corp.	0.1%	Financials
OAT	Quaker Oats Co.	0.1%	Consumer Staples
WB	Wachovia Corp. New	0.1%	Financials
BHI	Baker Hughes Inc.	0.1%	Energy
GIS	General Mls Inc.	0.1%	Consumer Staples
SLR	Solectron Corp.	0.1%	Capital Goods
TRB	Tribune Co. New	0.1%	Consumer Cyclicals
GX	Global Crossing Ltd	0.1%	Communication Services
SLM	USA Ed Inc.	0.1%	Financials
LTR	Loews Corp.	0.1%	Financials
MXIM	Maxim Integrated Prods Inc.	0.1%	Technology
FDX	Fedex Corp.	0.1%	Transportation
MHP	McGraw Hill Cos Inc.	0.1%	Consumer Cyclicals
EK	Eastman Kodak Co.	0.1%	Technology
BNI	Burlington Northn Santa Fe	0.1%	Transportation
XLNX	Xilinx Inc.	0.1%	Technology
SEBL	Siebel Sys Inc.	0.1%	Technology
HDI	Harley Davidson Inc.	0.1%	Consumer Cyclicals
S	Sears Roebuck & Co.	0.1%	Consumer Cyclicals
AL	ALCAN Inc.	0.1%	Basic Materials
CD	Cendant Corp.	0.1%	Consumer Cyclicals
K	Kellogg Co.	0.1%	Consumer Staples
NKE	Nike Inc.	0.1%	Consumer Cyclicals
COF	Capital One Finl Corp.	0.1%	Financials
KEY	Keycorp New	0.1%	Financials
NXTL	Nextel Communications Inc.	0.1%	Communication Services
WWY	Wrigley Wm Jr Co.	0.1%	Consumer Staples
CMA	Comerica Inc.	0.1%	Financials
MAS	Masco Corp.	0.1%	Consumer Cyclicals
WY	Weyerhaeuser Co.	0.1%	Basic Materials
MAY	May Dept Stores Co.	0.1%	Consumer Cyclicals
IPG	Interpublic Group Cos Inc.	0.1%	Consumer Cyclicals
FPL	FPL Group Inc.	0.1%	Utilities
TXU	TXU Corp.	0.1%	Utilities

(continued)

TABLE 5.2 (CONTINUED)
Companies in the Diversified Core Account

Ticker	Description	% Allocation	Sector
FRX	Forest Labs Inc.	0.1%	Health Care
GDW	Golden West Finl Corp. Del	0.1%	Financials
XEL	Xcel Energy Inc.	0.1%	Utilities
RTN/B	Raytheon Co.	0.1%	Technology
SYK	Stryker Corp.	0.1%	Health Care
MAR	Marriott Intl. Inc. New	0.1%	Consumer Cyclicals
CAG	Conagra Inc.	0.1%	Consumer Staples
AGN	Allergan Inc.	0.1%	Health Care
CMVT	Comverse Technology Inc.	0.1%	Technology
BEN	Franklin Res Inc.	0.1%	Financials
AVP	Avon Prods Inc.	0.1%	Consumer Staples
BR	Burlington Res Inc.	0.1%	Energy
RAL	Ralston Purina Co.	0.1%	Consumer Staples
AZA	Alza Corp.	0.1%	Health Care
BGEN	Biogen Inc.	0.1%	Health Care
SPC	St Paul Cos Inc.	0.1%	Financials
PEG	Public Svc Enterprise Group	0.1%	Utilities
OXY	Occidental Pete Corp.	0.1%	Energy
AOC	Aon Corp.	0.1%	Financials
HSY	Hershey Foods Corp.	0.1%	Consumer Staples
BDX	Becton Dickinson & Co.	0.1%	Health Care
APD	Air Prods & Chems Inc.	0.1%	Basic Materials
CEFT	Concord Efs Inc.	0.1%	Technology
YHOO	Yahoo Inc.	0.1%	Technology
PGN	Progress Energy Inc.	0.1%	Utilities
TJX	TJX Cos Inc. New	0.1%	Consumer Cyclicals
DE	Deere & Co.	0.1%	Capital Goods
MRO	USX Marathon Group	0.1%	Energy
UCL	Unocal Corp.	0.1%	Energy
ALTR	Altera Corp.	0.1%	Technology
PBI	Pitney Bowes Inc.	0.1%	Capital Goods
ADM	Archer Daniels Midland Co.	0.1%	Basic Materials
CHIR	Chiron Corp.	0.1%	Health Care
FD	Federated Dept Stores Inc. Del	0.1%	Consumer Cyclicals
ETR	Entergy Corp.	0.1%	Utilities
BSX	Boston Scientific Corp.	0.1%	Health Care
AMD	Advanced Micro Devices Inc.	0.1%	Technology
ADBE	Adobe Sys Inc.	0.1%	Technology
LNC	Lincoln Natl Corp. In	0.1%	Financials
DPH	Delphi Automotive Sys Corp.	0.1%	Consumer Cyclicals
MBI	MBIA Inc.	0.1%	Financials
ED	Consolidated Edison Inc.	0.1%	Utilities
SBUX	Starbucks Corp.	0.1%	Consumer Staples

TABLE 5.2 (CONTINUED)
Companies in the Diversified Core Account

Ticker	Description	% Allocation	Sector
DHR	Danaher Corp.	0.1%	Capital Goods
SOTR	Southtrust Corp.	0.1%	Financials
SNV	Synovus Finl Corp.	0.1%	Financials
CIT	CIT Group Inc.	0.1%	Financials
MAT	Mattel Inc.	0.1%	Consumer Cyclicals
TXT	Textron Inc.	0.1%	Capital Goods
UVN	Univision Communications Inc.	0.1%	Consumer Staples
MCK	McKesson Hboc Inc.	0.1%	Consumer Staples
NBR	Nabors Industries Inc.	0.1%	Energy
MEDI	Medimmune Inc.	0.1%	Health Care
DVN	Devon Energy Corporation New	0.1%	Energy
CCE	Coca Cola Enterprises Inc.	0.1%	Consumer Staples
BBY	Best Buy Co. Inc.	0.1%	Consumer Cyclicals
PPG	PPG Inds Inc.	0.1%	Basic Materials
CLX	Clorox Co.	0.1%	Consumer Staples
KLAC	KLA Tencor Corp.	0.1%	Technology
DOV	Dover Corp.	0.1%	Capital Goods
MTG	MGIC Invt Corp. Wis	0.1%	Financials
AAPL	Apple Computer	0.1%	Technology
PGR	Progressive Corp. Ohio	0.1%	Financials
NWL	Newell Rubbermaid Inc.	0.1%	Consumer Staples
PX	Praxair Inc.	0.1%	Basic Materials
CSX	CSX Corp.	0.1%	Transportation
APA	Apache Corp.	0.1%	Energy
UNM	Unumprovident Corp.	0.1%	Financials
JP	Jefferson Pilot Corp.	0.1%	Financials
BMET	Biomet Inc.	0.1%	Health Care
BBBY	Bed Bath & Beyond Inc.	0.1%	Consumer Cyclicals
MOLX	Molex Inc.	0.1%	Capital Goods
RX	IMS Health Inc.	0.1%	Consumer Cyclicals
RSH	Radioshack Corp.	0.1%	Consumer Cyclicals
SPLS	Staples Inc.	0.1%	Consumer Cyclicals
DG	Dollar Gen Corp.	0.1%	Consumer Cyclicals
ROH	Rohm & Haas Co.	0.1%	Basic Materials
ROK	Rockwell Intl. Corp. New	0.1%	Capital Goods
BRCM	Broadcom Corp.	0.1%	Technology
KG	King Pharmaceuticals Inc.	0.1%	Health Care
GP	Georgia Pac Corp.	0.1%	Basic Materials
SFA	Scientific Atlanta Inc.	0.1%	Technology
CEG	Constellation Energy Group Inc.	0.1%	Utilities
NYT	New York Times Co.	0.1%	Consumer Cyclicals
ADCT	ADC Telecommunications Inc.	0.1%	Technology

(continued)

TABLE 5.2 (CONTINUED)
Companies in the Diversified Core Account

Ticker	Description	% Allocation	Sector
ABK	Ambac Finl Group Inc.	0.1%	Financials
PSFT	Peoplesoft Inc.	0.1%	Technology
CTAS	Cintas Corp.	0.1%	Consumer Cyclicals
AHC	Amerada Hess Corp.	0.1%	Energy
IR	Ingersoll Rand Co.	0.1%	Capital Goods
HOT	Starwood Hotels & Resorts	0.1%	Consumer Cyclicals
NSC	Norfolk Southn Corp.	0.1%	Transportation
PPL	PPL Corp.	0.1%	Utilities
LTD	Limited Inc.	0.1%	Consumer Cyclicals
ASO	Amsouth Bancorporation	0.1%	Financials
RGBK	Regions Finl Corp.	0.1%	Financials
TOS	Tosco Corp.	0.1%	Energy
NOC	Northrop Grumman Corp.	0.1%	Capital Goods
KMI	Kinder Morgan Inc. Kans	0.1%	Utilities
NE	Noble Drilling Corp.	0.1%	Energy
KMG	Kerr McGee Corp.	0.1%	Energy
CINF	Cincinnati Finl Corp.	0.1%	Financials
NI	NiSource Inc.	0.1%	Utilities
SV	Stilwell Financial Inc.	0.1%	Financials
TSG	Sabre Hldgs Corp.	0.1%	Technology
WLP	Wellpoint Health Networks Inc.	0.1%	Health Care
FE	Firstenergy Corp.	0.1%	Utilities
SANM	Sanmina Corp.	0.1%	Capital Goods
AVY	Avery Dennison Corp.	0.1%	Capital Goods
FISV	Fiserv Inc.	0.1%	Technology
ABI	Applera Corp. Applied Biosys	0.1%	Health Care
TER	Teradyne Inc.	0.1%	Technology
DTE	DTE Energy Co.	0.1%	Utilities
CF	Charter One Finl Inc.	0.1%	Financials
ABX	Barrick Gold Corp.	0.1%	Basic Materials
YUM	Tricon Global Restaurants Inc.	0.1%	Consumer Staples
CCR	Countrywide Cr Inds Inc.	0.1%	Financials
CVG	Convergys Corp.	0.1%	Consumer Cyclicals
INTU	Intuit	0.1%	Technology
AEE	Ameren Corp.	0.1%	Utilities
LXK	Lexmark Intl. Inc.	0.1%	Technology
FO	Fortune Brands Inc.	0.1%	Consumer Staples
GTW	Gateway Inc.	0.1%	Technology
JCI	Johnson Ctls Inc.	0.1%	Capital Goods
ECL	Ecolab Inc.	0.1%	Basic Materials
CSC	Computer Sciences Corp.	0.1%	Technology
CIN	Cinergy Corp.	0.1%	Utilities
WPI	Watson Pharmaceuticals Inc.	0.1%	Health Care

TABLE 5.2 (CONTINUED)
Companies in the Diversified Core Account

Ticker	Description	% Allocation	Sector
NVLS	Novellus Sys Inc.	0.1%	Technology
BMC	Bmc Software Inc.	0.1%	Technology
AMR	AMR Corp. Del	0.1%	Transportation
OK	Old Kent Finl Corp.	0.1%	Financials
NTAP	Network Appliance Inc.	0.1%	Technology
KSE	Keyspan Corp.	0.1%	Utilities
AET	Aetna Inc.	0.1%	Health Care
UPC	Union Planters Corp.	0.1%	Financials
CNC	Conseco Inc.	0.1%	Financials
AYE	Allegheny Energy Inc.	0.1%	Utilities
SBL	Symbol Technologies Inc.	0.1%	Capital Goods
TMK	Torchmark Inc.	0.1%	Financials
WLL	Willamette Inds Inc.	0.1%	Basic Materials
BSC	Bear Stearns Cos Inc.	0.1%	Financials
LSI	Lsi Logic Corp.	0.1%	Technology
DAL	Delta Air Lines Inc. De	0.1%	Transportation
AMCC	Applied Micro Circuits Corp.	0.1%	Technology
PCG	PG&E Corp.	0.1%	Utilities
HRC	Healthsouth Corp.	0.1%	Health Care
EOG	EOG Resources Inc.	0.1%	Energy
SRE	Sempra Energy	0.1%	Utilities
UST	UST Inc.	0.1%	Consumer Staples
ETN	Eaton Corp.	0.1%	Capital Goods
NSM	National Semiconductor Corp.	0.1%	Technology
VMC	Vulcan Matls Co.	0.1%	Basic Materials
PALM	Palm Inc.	0.1%	Technology
HRB	Block H & R Inc.	0.0%	Consumer Cyclicals
GPC	Genuine Parts Co.	0.0%	Consumer Cyclicals
DJ	Dow Jones & Co. Inc.	0.0%	Consumer Cyclicals
TOY	Toys R Us Inc.	0.0%	Consumer Cyclicals
KM	K Mart Corp.	0.0%	Consumer Cyclicals
PH	Parker Hannifin Corp.	0.0%	Capital Goods
MCO	Moodys Corp.	0.0%	Financials
EFX	Equifax Inc.	0.0%	Technology
STJ	St Jude Med Inc.	0.0%	Health Care
BF/B	Brown Forman Corp.	0.0%	Consumer Staples
UIS	Unisys Corp.	0.0%	Technology
JCP	Penney J C Inc.	0.0%	Consumer Cyclicals
SHW	Sherwin Williams Co.	0.0%	Consumer Cyclicals
TRW	TRW Inc.	0.0%	Consumer Cyclicals
EIX	Edison Intl.	0.0%	Utilities
VTSS	Vitesse Semiconductor Corp.	0.0%	Technology

(continued)

TABLE 5.2 (CONTINUED)
Companies in the Diversified Core Account

Ticker	Description	% Allocation	Sector
JBL	Jabil Circuit Inc.	0.0%	Capital Goods
KRI	Knight Ridder Inc.	0.0%	Consumer Cyclicals
H	Harcourt Gen Inc.	0.0%	Consumer Cyclicals
RHI	Robert Half Intl. Inc.	0.0%	Consumer Staples
VFC	V F Corp.	0.0%	Consumer Cyclicals
XRX	Xerox Corp.	0.0%	Technology
GPU	GPU Inc.	0.0%	Utilities
TIF	Tiffany & Co. New	0.0%	Consumer Cyclicals
CTL	CenturyTel Inc.	0.0%	Communication Services
TMO	Thermo Electron Corp.	0.0%	Capital Goods
CTXS	Citrix Sys Inc.	0.0%	Technology
WIN	Winn Dixie Stores Inc.	0.0%	Consumer Staples
GR	Goodrich B F Co.	0.0%	Capital Goods
SIAL	Sigma Aldrich	0.0%	Basic Materials
TROW	Price T Rowe Group Inc.	0.0%	Financials
LEG	Leggett & Platt Inc.	0.0%	Consumer Cyclicals
PNW	Pinnacle West Cap Corp.	0.0%	Utilities
EMN	Eastman Chem Co.	0.0%	Basic Materials
HLT	Hilton Hotels Corp.	0.0%	Consumer Cyclicals
GT	Goodyear Tire And Rubber	0.0%	Consumer Cyclicals
NCR	NCR Corp. New	0.0%	Technology
AV	Avaya Inc.	0.0%	Technology
CMS	Cms Energy Corp.	0.0%	Utilities
CPWR	Compuware Corp.	0.0%	Technology
HBAN	Huntington Bancshares Inc.	0.0%	Financials
SAFC	Safeco Corp.	0.0%	Financials
HET	Harrahs Entmt Inc.	0.0%	Consumer Cyclicals
ITT	ITT Inds Inc.	0.0%	Capital Goods
FLR	Fluor Corp. New	0.0%	Capital Goods
PCAR	Paccar Inc.	0.0%	Capital Goods
MERQ	Mercury Interactive Corp.	0.0%	Technology
CZN	Citizens Communications Co.	0.0%	Communication Services
EC	Engelhard Corp.	0.0%	Basic Materials
NUE	Nucor Corp.	0.0%	Basic Materials
DNY	Donnelley R R & Sons Co.	0.0%	Consumer Staples
PD	Phelps Dodge Corp.	0.0%	Basic Materials
AZO	Autozone Inc.	0.0%	Consumer Cyclicals
CBE	Cooper Inds Inc.	0.0%	Capital Goods
GWW	Grainger W W Inc.	0.0%	Technology
NEM	Newmont Mng Corp.	0.0%	Basic Materials
AW	Allied Waste Industries Inc.	0.0%	Capital Goods
BDK	Black & Decker Corporation	0.0%	Consumer Cyclicals
WHR	Whirlpool Corp.	0.0%	Consumer Cyclicals

TABLE 5.2 (CONTINUED)
Companies in the Diversified Core Account

Ticker	Description	% Allocation	Sector
DRI	Darden Restaurants Inc.	0.0%	Consumer Staples
SWK	Stanley Works	0.0%	Consumer Cyclicals
PDG	Placer Dome Inc.	0.0%	Basic Materials
SEE	Sealed Air Corp. New	0.0%	Capital Goods
SUN	Sunoco Inc.	0.0%	Energy
ODP	Office Depot Inc.	0.0%	Consumer Cyclicals
ASH	Ashland Inc.	0.0%	Energy
NMK	Niagara Mohawk Hldgs Inc.	0.0%	Utilities
PLL	Pall Corp.	0.0%	Capital Goods
N	Inc.o Ltd	0.0%	Basic Materials
TEK	Tektronix Inc.	0.0%	Technology
RDC	Rowan Cos Inc.	0.0%	Energy
PKI	Perkinelmer Inc.	0.0%	Technology
MEA	Mead Corp.	0.0%	Basic Materials
DCN	Dana Corp.	0.0%	Consumer Cyclicals
APCC	American Pwr Conversion Corp.	0.0%	Capital Goods
LIZ	Liz Claiborne Inc.	0.0%	Consumer Cyclicals
WEN	Wendys Intl. Inc.	0.0%	Consumer Staples
MYG	Maytag Corp.	0.0%	Consumer Cyclicals
W	Westvaco Corp.	0.0%	Basic Materials
PMTC	Parametric Technology Corp.	0.0%	Technology
RKY	Coors Adolph Co.	0.0%	Consumer Staples
CTX	Centex Corp.	0.0%	Consumer Cyclicals
U	US Airways Group Inc.	0.0%	Transportation
CS	Cabletron Systems Inc.	0.0%	Technology
BOL	Bausch & Lomb Inc.	0.0%	Health Care
BCR	Bard C R Inc.	0.0%	Health Care
TIN	Temple Inland Inc.	0.0%	Basic Materials
FMC	FMC Corp.	0.0%	Basic Materials
HAS	Hasbro Inc.	0.0%	Consumer Cyclicals
IFF	International Flavors	0.0%	Basic Materials
CC	Circuit City Stores Inc.	0.0%	Consumer Cyclicals
ACV	Alberto Culver Co.	0.0%	Consumer Staples
MIL	Millipore Corp.	0.0%	Capital Goods
JWN	Nordstrom Inc.	0.0%	Consumer Cyclicals
QTRN	Quintiles Transnational Corp.	0.0%	Health Care
DDS	Dillards Inc.	0.0%	Consumer Cyclicals
HCR	Manor Care Inc. New	0.0%	Health Care
CNXT	Conexant Sys Inc.	0.0%	Technology
VC	Visteon Corp.	0.0%	Consumer Cyclicals
QLGC	Qlogic Corp.	0.0%	Technology
FCX	Freeport Mcmoran Copper & Gold	0.0%	Basic Materials

(continued)

TABLE 5.2 (CONTINUED)
Companies in the Diversified Core Account

Ticker	Description	% Allocation	Sector
PTV	Pactiv Corp.	0.0%	Basic Materials
BCC	Boise Cascade Corp.	0.0%	Basic Materials
ADSK	Autodesk Inc.orporated	0.0%	Technology
BMS	Bemis Inc.	0.0%	Basic Materials
MDP	Meredith Corp.	0.0%	Consumer Cyclicals
SVU	Supervalu Inc.	0.0%	Consumer Staples
BC	Brunswick Corp.	0.0%	Consumer Cyclicals
DLX	Deluxe Corp.	0.0%	Consumer Staples
HUM	Humana Inc.	0.0%	Health Care
NOVL	Novell Inc.	0.0%	Technology
GAS	Nicor Inc.	0.0%	Utilities
SNA	Snap On Inc.	0.0%	Consumer Cyclicals
CUM	Cummins Engine Inc.	0.0%	Capital Goods
GLK	Great Lakes Chemical Corp.	0.0%	Basic Materials
CR	Crane Co.	0.0%	Capital Goods
PHM	Pulte Corp.	0.0%	Consumer Cyclicals
KBH	KB Home	0.0%	Consumer Cyclicals
ATI	Allegheny Technologies Inc.	0.0%	Basic Materials
BVSN	Broadvision Inc.	0.0%	Technology
HPC	Hercules Inc.	0.0%	Basic Materials
HM	Homestake Mng Co.	0.0%	Basic Materials
NAV	Navistar Intl. Corp. Inc.	0.0%	Capital Goods
PGL	Peoples Energy Corp.	0.0%	Utilities
BLL	Ball Corp.	0.0%	Capital Goods
RBK	Reebok Intl. Ltd	0.0%	Consumer Cyclicals
X	USX U S Stl Group	0.0%	Basic Materials
TUP	Tupperware Corp.	0.0%	Consumer Staples
OKE	Oneok Inc. New	0.0%	Utilities
ANDW	Andrew Corp.	0.0%	Technology
CNS	Consolidated Stores Corp.	0.0%	Consumer Cyclicals
R	Ryder Sys Inc.	0.0%	Transportation
LDG	Longs Drug Stores Corp.	0.0%	Consumer Staples
PWER	Power One Inc.	0.0%	Capital Goods
TNB	Thomas & Betts Corp.	0.0%	Capital Goods
LPX	Louisiana Pac Corp.	0.0%	Basic Materials
TKR	Timken Co.	0.0%	Capital Goods
NSI	National Svc Inds Inc.	0.0%	Capital Goods
ADPT	Adaptec Inc.	0.0%	Technology
PCH	Potlatch Corp.	0.0%	Basic Materials
SAPE	Sapient Corp.	0.0%	Technology
BGG	Briggs & Stratton Corp.	0.0%	Capital Goods
CTB	Cooper Tire & Rubr Co.	0.0%	Consumer Cyclicals
WOR	Worthington Inds In	0.0%	Basic Materials

TABLE 5.2 (CONTINUED)
Companies in the Diversified Core Account

Ticker	Description	% Allocation	Sector
MDR	McDermott Intl. Inc.	0.0%	Capital Goods
AM	American Greetings Corp.	0.0%	Consumer Cyclicals
FSR	Firstar Corp. Wis	0.0%	Financials
MidCap SPDR S&P 400 (MDY)			
GENZ	Genzyme Corp.	1.11%	Health Care
ERTS	Electronic Arts	0.94%	Technology
MTB	M&T Bank Corp.	0.88%	Financials
SDS	SunGard Data Systems	0.85%	Technology
MLNM	Millennium Pharmaceuticals	0.84%	Health Care
DST	DST Systems Inc.	0.78%	Technology
WAT	Waters Corporation	0.78%	Technology
BJS	BJ Services	0.76%	Energy
IDPH	IDEC Pharmaceuticals	0.74%	Health Care
RJR	RJ Reynolds Tobacco Hldgs	0.74%	Consumer Staples
TDS	Telephone & Data Systems	0.71%	Communication Services
WPO	Washington Post	0.71%	Consumer Cyclicals
MI	Marshall & Ilsley Corp.	0.70%	Financials
WFT	Weatherford Int'l Inc.	0.70%	Energy
NCBC	National Commerce Bancorp	0.66%	Financials
IVX	IVAX Corp.	0.65%	Health Care
ESV	ENSCO Int'l	0.63%	Energy
ATML	Atmel Corp.	0.59%	Technology
CDN	Cadence Design Systems	0.59%	Technology
ZION	Zions Bancorp	0.59%	Financials
JNY	Jones Apparel Group	0.59%	Consumer Cyclicals
GLM	Global Marine	0.58%	Energy
NVDA	NVIDIA Corp.	0.57%	Technology
FDO	Family Dollar Stores	0.57%	Consumer Cyclicals
NFB	North Fork Bancorp	0.55%	Financials
BRW	Broadwing Inc.	0.54%	Communication Services
DGX	Quest Diagnostics	0.53%	Health Care
ASD	American Standard Cos	0.53%	Capital Goods
TE	TECO Energy	0.52%	Utilities
FTN	First Tennessee National	0.52%	Financials
HMA	Health Management Assoc	0.49%	Health Care
APOL	Apollo Group	0.49%	Consumer Cyclicals
IGT	International Game Technology	0.48%	Consumer Cyclicals
LIT	Litton Industries	0.48%	Technology
RATL	Rational Software	0.47%	Technology
DPL	DPL Inc.orporated	0.47%	Utilities
UCU	UtiliCorp. United	0.46%	Utilities

(continued)

TABLE 5.2 (CONTINUED)
Companies in the Diversified Core Account

Ticker	Description	% Allocation	Sector
DME	Dime Bancorp Inc.	0.46%	Financials
SII	Smith International	0.46%	Energy
BJ	BJ's Wholesale Club	0.45%	Consumer Cyclicals
SEIC	SEI Corp.	0.44%	Financials
ESRX	Express Scripts 'A'	0.44%	Health Care
ORI	Old Republic Int'l	0.43%	Financials
MCHP	Microchip Technology	0.43%	Technology
GPT	Greenpoint Financial Corp.	0.43%	Financials
ACS	Affiliated Computer Svcs	0.42%	Technology
ANF	Abercrombie & Fitch Co.	0.42%	Consumer Cyclicals
MYL	Mylan Laboratories	0.42%	Health Care
AWK	American Water Works	0.41%	Utilities
IDTI	Integrated Devices Tech	0.41%	Technology
SYMC	Symantec Corp.	0.41%	Technology
UDS	Ultramar Diamond Shamrock	0.41%	Energy
RDN	Radian Group	0.41%	Financials
BSYS	BISYS Group	0.40%	Technology
GILD	Gilead Sciences	0.40%	Health Care
RE	Everest Re Group	0.40%	Financials
PPE	Park Place Entertainment	0.40%	Consumer Cyclicals
TCB	TCF Financial	0.39%	Financials
NEU	Neuberger Berman Inc.	0.39%	Financials
HB	Hillenbrand Industries	0.39%	Capital Goods
MUR	Murphy Oil	0.39%	Energy
TSN	Tyson Foods	0.39%	Consumer Staples
AGE	Edwards (AG), Inc.	0.39%	Financials
CAM	Cooper Cameron Corp.	0.38%	Energy
LRCX	Lam Research	0.38%	Technology
BKNG	Banknorth Group Inc.	0.37%	Financials
MKC	McCormick & Co.	0.37%	Consumer Staples
PMI	PMI Group Inc.	0.37%	Financials
SNPS	Synopsys Inc.	0.37%	Technology
SCG	SCANA Corp. (New)	0.37%	Utilities
RDA	Readers Digest Assoc	0.37%	Consumer Cyclicals
NOI	National-Oilwell Inc.	0.36%	Energy
OEI	Ocean Energy Inc.(New)	0.36%	Energy
EAT	Brinker International	0.36%	Consumer Staples
SPW	SPX Corp.	0.36%	Capital Goods
LNCR	Lincare Holdings	0.36%	Health Care
AFC	Allmerica Financial	0.36%	Financials
VSH	Vishay Intertechnology	0.36%	Capital Goods
CDWC	CDW Computer Centers	0.35%	Consumer Cyclicals
HRL	Hormel Foods Corp.	0.35%	Consumer Staples

TABLE 5.2 (CONTINUED)
Companies in the Diversified Core Account

Ticker	Description	% Allocation	Sector
ASFC	Astoria Financial	0.35%	Financials
SCI	SCI Systems Inc.	0.34%	Capital Goods
LLL	L-3 Communications	0.34%	Technology
OXHP	Oxford Health Plans	0.34%	Health Care
LM	Legg Mason	0.34%	Financials
WEC	Wisconsin Energy	0.34%	Utilities
POM	Potomac Electric Power	0.34%	Utilities
ACF	AmeriCredit Corp.	0.34%	Financials
EXPD	Expeditors Int'l	0.34%	Transportation
MRBK	Mercantile Bankshares	0.34%	Financials
CBSS	Compass Bancshares	0.34%	Financials
AAS	AmeriSource Health 'A'	0.33%	Consumer Staples
TDW	Tidewater Inc.	0.33%	Energy
HNT	Health Net Inc.	0.33%	Health Care
IRF	Int'l Rectifier	0.33%	Technology
WON	Westwood One, Inc.	0.33%	Consumer Staples
LEN	Lennar Corp.	0.33%	Consumer Cyclicals
LNT	Alliant Energy	0.33%	Utilities
NU	Northeast Utilities	0.32%	Utilities
SEPR	Sepracor Inc.	0.32%	Health Care
PAS	PepsiAmericas, Inc.	0.32%	Consumer Staples
UNIT	Unitrin, Inc.	0.32%	Financials
MCRL	Micrel Inc.	0.31%	Technology
BOW	Bowater Inc.	0.31%	Basic Materials
WDR	Waddell & Reed Financial	0.31%	Financials
MDU	MDU Resources	0.30%	Utilities
NBL	Noble Affiliates	0.30%	Energy
CY	Cypress Semiconductor	0.30%	Technology
MCN	MCN Energy Group Inc.	0.30%	Utilities
BEC	Beckman Coulter Inc.	0.30%	Health Care
HP	Helmerich & Payne	0.30%	Energy
CHRW	CH Robinson Worldwide	0.30%	Transportation
TGP	The Timber Co.	0.30%	Basic Materials
ENR	Energizer Holdings Inc.	0.30%	Consumer Staples
FMER	FirstMerit Corp.	0.30%	Financials
TR	Tootsie Roll	0.29%	Consumer Staples
EQT	Equitable Resources	0.29%	Utilities
CKFR	CheckFree Corp.	0.29%	Technology
BBC	Bergen Brunswig	0.29%	Consumer Staples
ET	E*Trade Group	0.29%	Financials
ASBC	Associated Banc-Corp.	0.29%	Financials
ARW	Arrow Electronics	0.29%	Technology

(*continued*)

TABLE 5.2 (CONTINUED)
Companies in the Diversified Core Account

Ticker	Description	% Allocation	Sector
HIB	Hibernia Corp.	0.29%	Financials
CCN	Chris-Craft Industries	0.29%	Consumer Staples
STR	Questar Corp.	0.29%	Utilities
VVI	Viad Corp.	0.29%	Consumer Cyclicals
MAN	Manpower Inc.	0.28%	Consumer Staples
AJG	Gallagher (Arthur J)	0.28%	Financials
VLO	Valero Energy	0.28%	Energy
CSGS	CSG Systems Int'l	0.28%	Technology
VRTX	Vertex Pharmaceuticals	0.28%	Health Care
MVSN	Macrovision Corp.	0.28%	Technology
SON	Sonoco Products	0.28%	Basic Materials
AOT	Apogent Technologies	0.28%	Health Care
NFG	National Fuel Gas	0.27%	Utilities
DV	DeVRY Inc.	0.27%	Consumer Cyclicals
CBT	Cabot Corp.	0.27%	Basic Materials
FHCC	First Health Group Inc.	0.27%	Health Care
JKHY	Henry (Jack) & Assoc	0.27%	Technology
SVRN	Sovereign Bancorp	0.27%	Financials
HSP	Hispanic Broadcasting	0.27%	Consumer Staples
DLTR	Dollar Tree Stores	0.27%	Consumer Cyclicals
FAST	Fastenal Company	0.27%	Consumer Cyclicals
CPS	ChoicePoint Inc.	0.27%	Consumer Cyclicals
CEN	Ceridian Corp. (New)	0.27%	Consumer Cyclicals
EAS	Energy East	0.26%	Utilities
NST	NSTAR	0.26%	Utilities
GMT	GATX Corp.	0.26%	Transportation
ICN	ICN Pharmaceuticals	0.26%	Health Care
BRL	Barr Laboratories	0.26%	Health Care
AEOS	American Eagle Outfitters	0.26%	Consumer Cyclicals
FVB	First Virginia Banks	0.26%	Financials
MLM	Martin Marietta Materials	0.26%	Basic Materials
PL	Protective Life Corp.	0.26%	Financials
SMTC	Semtech Corp.	0.26%	Technology
OCR	Omnicare, Inc.	0.26%	Health Care
ISCA	International Speedway	0.26%	Consumer Cyclicals
DBD	Diebold, Inc.	0.25%	Capital Goods
PSD	Puget Energy, Inc. (Hldg Co.)	0.25%	Utilities
OSI	Outback Steakhouse	0.25%	Consumer Staples
LSCC	Lattice Semiconductor	0.25%	Technology
VRC	Varco International (New)	0.25%	Energy
COMS	3Com Corp.	0.25%	Technology
ALE	Allete	0.25%	Utilities
GLC	Galileo International	0.25%	Consumer Cyclicals

TABLE 5.2 (CONTINUED)
Companies in the Diversified Core Account

Ticker	Description	% Allocation	Sector
CIV	Conectiv	0.25%	Utilities
TGH	Trigon Healthcare Inc.	0.25%	Health Care
PDLI	Protein Design Labs	0.25%	Health Care
WL	Wilmington Trust Corp.	0.25%	Financials
RFMD	RF Micro Devices, Inc.	0.25%	Technology
TNO	True North Communications	0.25%	Consumer Cyclicals
AVT	Avnet, Inc.	0.25%	Technology
DNB	Dun & Bradstreet Corp. (New)	0.25%	Consumer Cyclicals
XRAY	Dentsply Int'l	0.24%	Health Care
MNMD	MiniMed Inc.	0.24%	Health Care
HC	Hanover Compressor Hldg Co.	0.24%	Energy
BLC	Belo Corp.	0.24%	Consumer Cyclicals
GRP	Grant Prideco	0.24%	Energy
LEA	Lear Corporation	0.24%	Consumer Cyclicals
ACXM	Acxiom Corp.	0.24%	Consumer Cyclicals
SKS	Saks Inc.orporated	0.24%	Consumer Cyclicals
CYN	City National Corp.	0.24%	Financials
POS	Catalina Marketing	0.23%	Consumer Cyclicals
LAB	LaBranche & Co.	0.23%	Financials
NDE	IndyMac Bancorp	0.23%	Financials
OGE	OGE Energy Corp.	0.23%	Utilities
LUK	Leucadia National Corp.	0.23%	Financials
ETM	Entercom Communications	0.23%	Consumer Staples
MLHR	Miller (Herman)	0.23%	Capital Goods
REY	Reynolds & Reynolds	0.23%	Capital Goods
IFIN	Investors Financial Services	0.23%	Financials
DSS	Quantum Corp. -DSSG Stock	0.23%	Technology
IBP	IBP, Inc.	0.22%	Consumer Staples
WIND	Wind River Systems	0.22%	Technology
NNS	Newport News Shipbuilding	0.22%	Capital Goods
GNTX	Gentex Corp.	0.22%	Consumer Cyclicals
PCP	Precision Castparts	0.22%	Capital Goods
LYO	Lyondell Chemical Co.	0.22%	Basic Materials
WR	Western Resources	0.22%	Utilities
MNY	The MONY Group	0.22%	Financials
CMH	Clayton Homes	0.22%	Consumer Cyclicals
NATI	National Instruments	0.21%	Technology
ELY	Callaway Golf Co.	0.21%	Consumer Cyclicals
DQE	DQE, Inc.	0.21%	Utilities
HRS	Harris Corp.	0.21%	Technology
ITG	Investment Technology Group (N	0.21%	Technology
DS	Dallas Semiconductor	0.21%	Technology

(continued)

TABLE 5.2 (CONTINUED)
Companies in the Diversified Core Account

Ticker	Description	% Allocation	Sector
AFG	American Financial Group Hldg	0.21%	Financials
TECD	Tech Data Corp.	0.21%	Consumer Cyclicals
LZ	Lubrizol Corp.	0.20%	Basic Materials
TFX	Teleflex	0.20%	Capital Goods
VCI	Valassis Communication	0.20%	Consumer Staples
NMG/A	Neiman-Marcus Group 'A'	0.20%	Consumer Cyclicals
PXD	Pioneer Natural Resources	0.20%	Energy
BKS	Barnes & Noble	0.20%	Consumer Cyclicals
JEC	Jacobs Engineering Group	0.20%	Capital Goods
MBG	Mandalay Resort Group	0.20%	Consumer Cyclicals
PKS	Six Flags Inc.	0.20%	Consumer Cyclicals
KLT	Kansas City Power & Light	0.20%	Utilities
HHS	Harte-Hanks, Inc.	0.20%	Consumer Cyclicals
ROST	Ross Stores	0.20%	Consumer Cyclicals
BOH	Pacific Century Financial Corp.	0.20%	Financials
IFMX	Informix Corp.	0.20%	Technology
MRL	Marine Drilling Co.	0.20%	Energy
SRM	Sensormatic Electronics	0.19%	Capital Goods
WSM	Williams-Sonoma Inc.	0.19%	Consumer Cyclicals
MTP	Montana Power	0.19%	Utilities
KEM	KEMET Corp.	0.19%	Capital Goods
MHK	Mohawk Industries	0.19%	Consumer Cyclicals
WBST	Webster Financial Corp.	0.19%	Financials
VVC	Vectren Corporation	0.19%	Utilities
IDA	IDACORP Inc. Hldg Co.	0.19%	Utilities
IGL	IMC Global Inc.	0.18%	Basic Materials
CNF	CNF Inc.	0.18%	Transportation
RSLN	Roslyn Bancorp	0.18%	Financials
HNI	HON Industries	0.18%	Capital Goods
HUB/B	Hubbell Inc. (Class B)	0.18%	Capital Goods
WABC	Westamerica Bancorp	0.18%	Financials
PFGI	Provident Financial Group Inc.	0.18%	Financials
SNDK	SanDisk Corp.	0.18%	Technology
PSS	Payless ShoeSource Inc. Hldg	0.18%	Consumer Cyclicals
SYBS	Sybase Inc.	0.18%	Technology
MENT	Mentor Graphics	0.17%	Technology
LEE	Lee Enterprises	0.17%	Consumer Cyclicals
HTN	Houghton Mifflin	0.17%	Consumer Cyclicals
BGP	Borders Group	0.17%	Consumer Cyclicals
SZA	Suiza Foods Corp.	0.17%	Consumer Staples
CYT	Cytec Industries	0.17%	Basic Materials
WGL	WGL Holdings, Inc.	0.17%	Utilities
PWR	Quanta Services	0.17%	Capital Goods

TABLE 5.2 (CONTINUED)
Companies in the Diversified Core Account

Ticker	Description	% Allocation	Sector
AHG	Apria Healthcare Group	0.16%	Health Care
CK	Crompton Corporation	0.16%	Basic Materials
SOI	Solutia Inc.	0.16%	Basic Materials
SCHL	Scholastic Corp.	0.16%	Consumer Cyclicals
PNR	Pentair Corp.	0.16%	Capital Goods
CORR	Cor Therapeutics	0.16%	Health Care
VAL	Valspar Corp.	0.16%	Basic Materials
HE	Hawaiian Electric Industries	0.16%	Utilities
CDO	Comdisco, Inc.	0.16%	Technology
NIS	NOVA Corp.	0.16%	Technology
DF	Dean Foods	0.16%	Consumer Staples
EMMS	Emmis Communications	0.15%	Consumer Staples
ATG	AGL Resources Ltd	0.15%	Utilities
DL	Dial Corp.	0.15%	Consumer Staples
DCI	Donaldson Co.	0.15%	Capital Goods
CRUS	Cirrus Logic	0.15%	Technology
FBN	Furniture Brands Int'l	0.15%	Consumer Cyclicals
SWFT	Swift Transportation	0.15%	Transportation
TQNT	TriQuint Semiconductor	0.15%	Technology
SRP	Sierra Pacific Resources (New)	0.15%	Utilities
AFCI	Advanced Fibre Communications	0.15%	Technology
EW	Edwards Lifesciences Corp.	0.15%	Health Care
SIVB	Silicon Valley Bancshares	0.15%	Financials
URI	United Rentals	0.15%	Consumer Cyclicals
EAGL	EGL Inc.	0.15%	Transportation
NETA	Network Associates Inc.	0.15%	Technology
PNM	Public Service of New Mexico	0.15%	Utilities
STK	Storage Technology	0.15%	Technology
CAR	Carter-Wallace	0.15%	Consumer Staples
PZB	Pittston Brink's Group	0.15%	Consumer Cyclicals
RYN	Rayonier Inc.	0.15%	Basic Materials
SXT	Sensient Technologies	0.14%	Consumer Staples
CMOS	Credence Systems	0.14%	Technology
PZL	Pennzoil-Quaker State (New)	0.14%	Energy
LANC	Lancaster Colony	0.14%	Consumer Cyclicals
QHGI	Quorom Health Group	0.14%	Health Care
BTH	Blyth Inc.	0.14%	Consumer Cyclicals
TXCC	TranSwitch Corp.	0.14%	Technology
BID	Sotheby's Holdings	0.14%	Consumer Cyclicals
AKS	AK Steel Hldg Corp.	0.14%	Basic Materials
CGO	Atlas Air Worldwide Hldgs	0.14%	Transportation
UVV	Universal Corp.	0.14%	Basic Materials

(*continued*)

TABLE 5.2 (CONTINUED)
Companies in the Diversified Core Account

Ticker	Description	% Allocation	Sector
GBBK	Greater Bay Bancorp	0.14%	Financials
PLXS	Plexus Corp.	0.14%	Capital Goods
CCMP	Cabot Microelectronics Corp.	0.14%	Technology
LGTO	Legato Systems Inc.	0.14%	Technology
YRK	York International	0.14%	Capital Goods
MEG/A	Media General	0.14%	Consumer Cyclicals
BWA	Borg Warner Inc.	0.14%	Consumer Cyclicals
BKH	Black Hills	0.14%	Utilities
NEWP	Newport Corporation	0.14%	Technology
EYE	VISX Inc.	0.14%	Health Care
RPM	RPM Inc.	0.14%	Basic Materials
ALB	Albemarle Corp.	0.13%	Basic Materials
CNL	Cleco Corp. Hldg Co.	0.13%	Utilities
CBRL	CBRL Group, Inc.	0.13%	Consumer Staples
PLCM	Polycom Inc.	0.13%	Technology
CSL	Carlisle Companies	0.13%	Capital Goods
INCY	INCYTE Genomics Inc.	0.13%	Health Care
HSC	Harsco Corp.	0.13%	Capital Goods
ADTN	Adtran Inc.	0.13%	Technology
TTN	Titan Corp.	0.13%	Technology
STE	STERIS Corp.	0.13%	Health Care
PR	Price Communications	0.12%	Communication Services
AVCT	Avocent Corp.	0.12%	Technology
MACR	Macromedia Inc.	0.12%	Technology
OSG	Overseas Shipholding Group	0.12%	Transportation
GVA	Granite Construction	0.12%	Capital Goods
GTK	GTECH Holdings Corp.	0.12%	Consumer Cyclicals
TECUA	Tecumseh Products Co.	0.12%	Capital Goods
ARM	ArvinMeritor Inc.	0.12%	Consumer Cyclicals
RETK	Retek Inc.	0.12%	Technology
DOL	Dole Foods	0.12%	Consumer Staples
OLN	Olin Corp.	0.12%	Basic Materials
KEA	Keane Inc.	0.12%	Technology
SUP	Superior Industries	0.12%	Consumer Cyclicals
MIPSB	MIPS Technologies	0.12%	Technology
AME	Ametek, Inc.	0.12%	Capital Goods
FSS	Federal Signal	0.11%	Capital Goods
PLT	Plantronics Inc.	0.11%	Technology
CLE	Claire's Stores	0.11%	Consumer Cyclicals
ALEX	Alexander & Baldwin	0.11%	Transportation
PWAV	Powerwave Technologies	0.11%	Communication Services
CTV	CommScope, Inc.	0.11%	Technology
PHSY	PacifiCare Health Sys	0.11%	Health Care

TABLE 5.2 (CONTINUED)
Companies in the Diversified Core Account

Ticker	Description	% Allocation	Sector
NDSN	Nordson Corporation	0.11%	Capital Goods
KELYA	Kelly Services	0.11%	Consumer Staples
FLS	Flowserve Corp.	0.11%	Capital Goods
CHD	Church & Dwight	0.11%	Consumer Staples
KMT	Kennametal Inc.	0.11%	Capital Goods
COV	Covanta Energy Corp.	0.11%	Consumer Cyclicals
KDN	Kaydon Corp.	0.10%	Capital Goods
LE	Lands' End	0.10%	Consumer Cyclicals
IMN	Imation Corp.	0.10%	Technology
IBC	Interstate Bakeries	0.10%	Consumer Staples
SLVN	Sylvan Learning Systems	0.10%	Consumer Cyclicals
SAWS	Sawtek Inc.	0.10%	Technology
MODI	Modine Mfg	0.10%	Consumer Cyclicals
CVD	Covance Inc.	0.10%	Health Care
WCLX	Wisconsin Central Trans	0.10%	Transportation
DRYR	Dreyer's Grand Ice Cream	0.09%	Consumer Staples
PRGO	Perrigo Co.	0.09%	Consumer Staples
TRN	Trinity Industries	0.09%	Capital Goods
HMN	Horace Mann Educators	0.09%	Financials
MTX	Minerals Technologies	0.09%	Basic Materials
FOE	Ferro Corp.	0.09%	Basic Materials
ALK	Alaska Air Group	0.09%	Transportation
LFB	Longview Fibre	0.09%	Basic Materials
WMO	Wausau-Mosinee Paper	0.09%	Basic Materials
USG	USG Corp.	0.09%	Consumer Cyclicals
WCS	Wallace Computer Services	0.09%	Capital Goods
NCOG	NCO Group Inc.	0.08%	Consumer Cyclicals
MTZ	MasTec Inc.	0.08%	Capital Goods
RDK	Ruddick Corp.	0.08%	Consumer Staples
SJM	The JM Smucker Co.	0.08%	Consumer Staples
INFS	InFocus Corp.	0.08%	Technology
KFY	Korn/Ferry International	0.08%	Consumer Staples
CRS	Carpenter Technology	0.08%	Basic Materials
BOBE	Bob Evans Farms	0.08%	Consumer Staples
SSSS	Stewart & Stevenson Services	0.08%	Capital Goods
BN	Banta Corp.	0.08%	Consumer Staples
FULL	Fuller (HB) Co.	0.08%	Basic Materials
ROL	Rollins, Inc.	0.08%	Consumer Cyclicals
AG	AGCO Corp.	0.07%	Capital Goods
AIN	Albany International	0.07%	Capital Goods
OCAS	Ohio Casualty	0.07%	Financials
BDG	Bandag Inc.	0.07%	Consumer Cyclicals

(*continued*)

TABLE 5.2 (CONTINUED)
Companies in the Diversified Core Account

Ticker	Description	% Allocation	Sector
JBHT	Hunt(JB) Transport Serv Inc.	0.07%	Transportation
PZZA	Papa John's Int'l	0.07%	Consumer Staples
IT/B	Gartner, Inc.	0.07%	Technology
DY	Dycom Industries	0.07%	Capital Goods
GLT	PH Glatfelter Co.	0.07%	Basic Materials
ARG	Airgas Inc.	0.07%	Basic Materials
UCR	UCAR International	0.07%	Basic Materials
SDRC	Structural Dynamics Research	0.07%	Technology
ABF	Airborne, Inc.	0.06%	Transportation
WXS	Westpoint Stevens	0.06%	Consumer Cyclicals
MPS	Modis Professional Svc	0.06%	Consumer Staples
DSPG	DSP Group	0.05%	Technology
SQA/A	Sequa Corp.	0.05%	Capital Goods
UFI	Unifi, Inc.	0.05%	Consumer Cyclicals
SHLM	Schulman (A), Inc.	0.05%	Basic Materials
LNCE	Lance, Inc.	0.04%	Consumer Staples
ANTC	ANTEC Corporation	0.04%	Technology
TSAI	Transaction Systems Architects	0.03%	Technology
NCH	NCH Corp.	0.03%	Consumer Cyclicals
RT	Ryerson Tull, Inc. (New)	0.03%	Capital Goods
STAR	Lone Star Steakhouse	0.03%	Consumer Staples
SYKE	Sykes Enterprises	0.03%	Technology
MXM	MAXXAM Inc.	0.01%	Basic Materials
NASDAQ 100 (QQQ)			
MSFT	Microsoft Corporation	9.5%	Technology
INTC	Intel Corporation	5.9%	Technology
QCOM	QUALCOMM Inc.orporated	4.9%	Technology
CSCO	Cisco Systems, Inc.	3.8%	Technology
ORCL	Oracle Corporation	3.3%	Technology
AMGN	Amgen Inc.	2.8%	Health Care
VSTR	VoiceStream Wireless Corporation	2.4%	Communication Services
JDSU	JDS Uniphase Corporation	2.3%	Technology
DELL	Dell Computer Corporation	2.3%	Technology
SUNW	Sun Microsystems, Inc.	2.1%	Technology
WCOM	WorldCom, Inc.	1.8%	Communication Services
CMCSK	Comcast Corporation	1.8%	Consumer Staples
VRTS	VERITAS Software Corporation	1.8%	Technology
AMAT	Applied Materials, Inc.	1.7%	Technology
LLTC	Linear Technology Corporation	1.6%	Technology
MXIM	Maxim Integrated Products, Inc.	1.6%	Technology
XLNX	Xilinx, Inc.	1.5%	Technology
CIEN	CIENA Corporation	1.4%	Technology

An Example: Thousand Stock Portfolio

TABLE 5.2 (CONTINUED)
Companies in the Diversified Core Account

Ticker	Description	% Allocation	Sector
SEBL	Siebel Systems, Inc.	1.3%	Technology
NXTL	Nextel Communications, Inc.	1.3%	Communication Services
PAYX	Paychex, Inc.	1.2%	Technology
GMST	Gemstar-TV Guide International Inc.	1.2%	Consumer Cyclicals
ALTR	Altera Corporation	1.1%	Technology
CHIR	Chiron Corporation	1.1%	Health Care
BGEN	Biogen, Inc.	1.1%	Health Care
SBUX	Starbucks Corporation	1.1%	Consumer Staples
CHKP	Check Point Software Technologies Ltd.	1.0%	Technology
AAPL	Apple Computer, Inc.	1.0%	Technology
IMNX	Immunex Corporation	1.0%	Health Care
BEAS	BEA Systems, Inc.	1.0%	Technology
CMVT	Comverse Technology, Inc.	1.0%	Technology
GENZ	Genzyme General	0.9%	Health Care
CEFT	Concord EFS, Inc.	0.9%	Technology
TLAB	Tellabs, Inc.	0.9%	Technology
BMET	Biomet, Inc.	0.9%	Health Care
PSFT	PeopleSoft, Inc.	0.9%	Technology
BBBY	Bed Bath & Beyond Inc.	0.9%	Consumer Cyclicals
COST	Costco Wholesale Corporation	0.9%	Consumer Cyclicals
KLAC	KLA-Tencor Corporation	0.8%	Technology
USAI	USA Networks, Inc.	0.8%	Consumer Staples
ADCT	ADC Telecommunications, Inc.	0.8%	Technology
ADBE	Adobe Systems Inc.orporated	0.8%	Technology
MEDI	MedImmune, Inc.	0.8%	Health Care
CTAS	Cintas Corporation	0.8%	Consumer Cyclicals
JNPR	Juniper Networks, Inc.	0.8%	Technology
SPOT	PanAmSat Corporation	0.8%	Technology
FLEX	Flextronics International Ltd.	0.7%	Capital Goods
ERTS	Electronic Arts Inc.	0.7%	Technology
INTU	Intuit Inc.	0.7%	Technology
EBAY	eBay Inc.	0.7%	Technology
MLNM	Millennium Pharmaceuticals, Inc.	0.7%	Health Care
FISV	Fiserv, Inc.	0.7%	Technology
DISH	EchoStar Communications Corporation	0.6%	Consumer Staples
VRSN	VeriSign, Inc.	0.6%	Technology
ADLAC	Adelphia Communications Corporation	0.6%	Consumer Staples
SANM	Sanmina Corporation	0.6%	Capital Goods
IDPH	IDEC Pharmaceuticals Corporation	0.6%	Health Care
EXDS	Exodus Communications, Inc.	0.6%	Technology
ITWO	i2 Technologies, Inc.	0.6%	Technology
NVLS	Novellus Systems, Inc.	0.6%	Technology

(continued)

TABLE 5.2 (CONTINUED)
Companies in the Diversified Core Account

Ticker	Description	% Allocation	Sector
AMCC	Applied Micro Circuits Corporation	0.5%	Technology
NTAP	Network Appliance, Inc.	0.5%	Technology
HGSI	Human Genome Sciences, Inc.	0.5%	Health Care
ERICY	LM Ericsson Telephone Company	0.5%	Technology
VTSS	Vitesse Semiconductor Corporation	0.5%	Technology
BRCD	Brocade Communications Systems, Inc.	0.5%	Technology
CTXS	Citrix Systems, Inc.	0.4%	Technology
SPLS	Staples, Inc.	0.4%	Consumer Cyclicals
PALM	Palm, Inc.	0.4%	Technology
LVLT	Level 3 Communications, Inc.	0.4%	Communication Services
PMCS	PMC - Sierra, Inc.	0.4%	Technology
YHOO	Yahoo! Inc.	0.4%	Technology
MCLD	McLeodUSA Inc.orporated	0.4%	Communication Services
BRCM	Broadcom Corporation	0.4%	Technology
TMPW	TMP Worldwide Inc.	0.4%	Consumer Cyclicals
PCAR	PACCAR Inc.	0.4%	Capital Goods
MERQ	Mercury Interactive Corporation	0.3%	Technology
RATL	Rational Software Corporation	0.3%	Technology
ATML	Atmel Corporation	0.3%	Technology
MOLX	Molex Inc.orporated	0.3%	Capital Goods
PMTC	Parametric Technology Corporation	0.3%	Technology
SSCC	Smurfit-Stone Container Corporation	0.3%	Basic Materials
MFNX	Metromedia Fiber Network, Inc.	0.3%	Technology
MCHP	Microchip Technology Inc.orporated	0.3%	Technology
AMZN	Amazon.com, Inc.	0.2%	Consumer Cyclicals
CNXT	Conexant Systems, Inc.	0.2%	Technology
CPWR	Compuware Corporation	0.2%	Technology
RFMD	RF Micro Devices, Inc.	0.2%	Technology
QLGC	QLogic Corporation	0.2%	Technology
XOXO	XO Communications, Inc.	0.2%	Communication Services
NOVL	Novell, Inc.	0.2%	Technology
ARBA	Ariba, Inc.	0.2%	Technology
ABGX	Abgenix, Inc.	0.2%	Health Care
CNET	CNET Networks, Inc.	0.2%	Technology
BVSN	BroadVision, Inc.	0.2%	Technology
ATHM	At Home Corporation	0.1%	Technology
RNWK	RealNetworks, Inc.	0.1%	Technology
COMS	3Com Corporation	0.1%	Technology
CMGI	CMGI, Inc.	0.1%	Technology
INKT	Inktomi Corporation	0.1%	Technology

TABLE 5.3
Companies in the Individual Active Account Sector SPDRs

Ticker	Description	% Allocation	Group
	Basic Industries Select Sector SPDR (XLB)		
DD	Du Pont E I De Nemours & Co.	16.14%	Chemicals
AA	Alcoa Inc.	11.82%	Aluminum
DOW	Dow Chem Co.	10.75%	Chemicals
IP	International Paper Co.	6.60%	Paper & Forest Products
AL	Alcan Inc.	4.35%	Aluminum
WY	Weyerhaeuser Co.	4.19%	Paper & Forest Products
APD	Air Prods & Chems Inc.	3.35%	Chemicals
PPG	PPG Inds Inc.	2.94%	Chemicals - Diversified
PX	Praxair Inc.	2.69%	Chemicals
ROH	Rohm & Haas Co.	2.57%	Chemicals
GP	Georgia Pac Corp.	2.52%	Paper & Forest Products
AVY	Avery Dennison Corp.	2.18%	Manufacturing (Specialized)
ABX	Barrick Gold Corp.	2.15%	Gold/Precious Metals Mining
ECL	Ecolab Inc.	2.05%	Chemicals Specialty
WLL	Willamette Inds Inc.	1.91%	Paper & Forest Products
VMC	Vulcan Matls Co.	1.80%	Construction
EMN	Eastman Chem Co.	1.44%	Chemicals
SIAL	Sigma Aldrich	1.41%	Chemicals Specialty
EC	Engelhard Corp.	1.26%	Chemicals - Diversified
PD	Phelps Dodge Corp.	1.20%	Metals Mining
NUE	Nucor Corp.	1.18%	Steel
NEM	Newmont Mng Corp.	1.18%	Gold/Precious Metals Mining
PDG	Placer Dome Inc.	1.08%	Gold/Precious Metals Mining
SEE	Sealed Air Corp. New	1.06%	Manufacturing (Specialized)
N	Inc.o Ltd	1.02%	Metals Mining
MEA	Mead Corp.	0.95%	Paper & Forest Products
W	Westvaco Corp.	0.93%	Paper & Forest Products
FMC	FMC Corp.	0.85%	Chemicals - Diversified
TIN	Temple Inland Inc.	0.82%	Containers/Packaging (Paper)
FCX	Freeport Mcmoran Copper & Gold	0.74%	Metals Mining
PTV	Pactiv Corp.	0.73%	Containers/Packaging (Paper)
BCC	Boise Cascade Corp.	0.68%	Paper & Forest Products
BMS	Bemis Inc.	0.67%	Containers/Packaging (Paper)
GLK	Great Lakes Chemical Corp.	0.59%	Chemicals Specialty
HPC	Hercules Inc.	0.53%	Chemicals Specialty
ATI	Allegheny Technologies Inc.	0.53%	Steel
HM	Homestake Mng Co.	0.53%	Gold/Precious Metals Mining
X	USX U S Stl Group	0.50%	Steel
BLL	Ball Corp.	0.49%	Containers - Metal & Glass
LPX	Louisiana Pac Corp.	0.38%	Paper & Forest Products
TKR	Timken Co.	0.35%	Machinery - Diversified

(continued)

TABLE 5.3 (CONTINUED)
Companies in the Individual Active Account Sector SPDRs

Ticker	Description	% Allocation	Group
PCH	Potlatch Corp.	0.34%	Paper & Forest Products
WOR	Worthington Inds In	0.31%	Steel
Consumer Services Select SPDR (XLV)			
VIA/B	Viacom Inc.	14.47%	Entertainment
DIS	Disney Walt Co.	11.22%	Entertainment
CCU	Clear Channel Communications	6.15%	Broadcasting (TV, Radio & Cable)
CMCSK	Comcast Corp.	6.14%	Broadcasting (TV, Radio & Cable)
HCA	HCA Healthcare Co.	4.34%	Health Care (Hospital Mgmt)
MCD	McDonalds Corp.	4.31%	Restaurants
UNH	UnitedHealth Group Inc.	3.74%	Health Care (Managed Care)
CCL	Carnival Corp.	3.21%	Lodging - Hotels
CI	Cigna Corp.	3.11%	Health Care (Managed Care)
GCI	Gannett Inc.	3.08%	Publishing - Newspapers
OMC	Omnicom Group	2.87%	Services (Advertising/Mktg)
THC	Tenet Healthcare Corp.	2.84%	Health Care (Hospital Mgmt)
TRB	Tribune Co. New	2.45%	Publishing - Newspapers
MHP	Mcgraw Hill Cos Inc.	2.32%	Publishing
CD	Cendant Corp.	2.25%	Services (Commercial & Consum)
IPG	Interpublic Group Cos Inc.	2.09%	Services (Advertising/Mktg)
MAR	Marriott Intl. Inc. New	1.99%	Lodging - Hotels
SBUX	Starbucks Corp.	1.64%	Restaurants
RX	IMS Health Inc.	1.56%	Services (Commercial & Consum)
UVN	Univision Communications Inc.	1.49%	Broadcasting (TV, Radio & Cable)
NYT	New York Times Co.	1.36%	Publishing - Newspapers
WLP	Wellpoint Health Networks Inc.	1.27%	Health Care (Managed Care)
CTAS	Cintas Corp.	1.26%	Services (Commercial & Consum)
HOT	Starwood Hotels & Resorts	1.24%	Lodging - Hotels
YUM	Tricon Global Restaurants Inc.	1.21%	Restaurants
HRC	Healthsouth Corp.	1.14%	Health Care (Special Serv)
HRB	Block H & R Inc.	1.02%	Services (Commercial & Consum)
AET	Aetna Inc.	0.97%	Health Care (Managed Care)
DJ	Dow Jones & Co. Inc.	0.94%	Publishing — Newspapers
HLT	Hilton Hotels Corp.	0.85%	Lodging — Hotels
KRI	Knight Ridder Inc.	0.85%	Publishing — Newspapers
HET	Harrahs Entmt Inc.	0.79%	Gaming, Lottery, & Parimutuel
RHI	Robert Half Intl. Inc.	0.75%	Services (Employment)
DNY	Donnelley R R & Sons Co.	0.73%	Speciality Printing
HCR	Manor Care Inc. New	0.73%	Health Care (Long-Term Care)
DRI	Darden Restaurants Inc.	0.69%	Restaurants
WEN	Wendys Intl. Inc.	0.62%	Restaurants
QTRN	Quintiles Transnational Corp.	0.57%	Health Care (Special Serv)
HUM	Humana Inc.	0.56%	Health Care (Managed Care)

TABLE 5.3 (CONTINUED)
Companies in the Individual Active Account Sector SPDRs

Ticker	Description	% Allocation	Group
MDP	Meredith Corp.	0.44%	Publishing
DLX	Deluxe Corp.	0.44%	Speciality Printing
AM	American Greetings Corp.	0.19%	Consumer (Jewelry/Novelties)
Consumer Staples Select Sector SPDR (XLP)			
PFE	Pfizer Inc.	11.78%	Health Care (Drugs/Pharms)
MRK	Merck & Co. Inc.	7.98%	Health Care (Drugs/Pharms)
JNJ	Johnson & Johnson	5.54%	Health Care Diversified
BMY	Bristol Myers Squibb Co.	5.30%	Health Care Diversified
KO	Coca Cola Co.	5.12%	Beverages (Non-Alcoholic)
MO	Philip Morris Cos Inc.	4.81%	Tobacco
LLY	Lilly Eli & Co.	3.94%	Health Care (Drugs/Pharms)
PG	Procter & Gamble Co.	3.71%	Household Prods (Non-Durable)
AHP	American Home Products Corp.	3.51%	Health Care Diversified
ABT	Abbott Labs	3.33%	Health Care Diversified
PHA	Pharmacia Corp.	2.96%	Health Care (Drugs/Pharms)
PEP	Pepsico Inc.	2.89%	Beverages (Non-Alcoholic)
AMGN	Amgen Inc.	2.86%	Biotechnology
MDT	Medtronic Inc.	2.50%	Health Care (Medical Prods/Sups)
SGP	Schering Plough Corp.	2.44%	Health Care (Drugs/Pharms)
WAG	Walgreen Co.	1.89%	Retail Stores — Drug Store
BUD	Anheuser Busch Cos Inc.	1.89%	Beverages — Alcoholic
KMB	Kimberly Clark Corp.	1.65%	Household Prods (Non-Durable)
G	Gillette Co.	1.50%	Personal Care
CL	Colgate Palmolive Co.	1.44%	Household Prods (Non-Durable)
UN	Unilever N V	1.37%	Foods
BAX	Baxter Intl. Inc.	1.26%	Health Care (Medical Prods/Sups)
SWY	Safeway Inc.	1.26%	Retail Stores — Food Chains
CAH	Cardinal Health Inc.	1.24%	Distributors (Food & Health)
CVS	CVS Corp.	1.05%	Retail Stores — Drug Store
KR	Kroger Co.	0.96%	Retail Stores — Food Chains
SYY	Sysco Corp.	0.82%	Distributors (Food & Health)
SLE	Sara Lee Corp.	0.81%	Foods
HNZ	Heinz H J Co.	0.64%	Foods
GDT	Guidant Corp.	0.63%	Health Care (Medical Prods/Sups)
ABS	Albertsons Inc.	0.59%	Retail Stores — Food Chains
OAT	Quaker Oats Co.	0.58%	Foods
CPB	Campbell Soup Co.	0.57%	Foods
GIS	General Mls Inc.	0.56%	Foods
K	Kellogg Co.	0.50%	Foods
WWY	Wrigley Wm Jr Co.	0.50%	Foods
FRX	Forest Labs Inc.	0.48%	Health Care (Drugs/Pharms)

(continued)

TABLE 5.3 (CONTINUED)
Companies in the Individual Active Account Sector SPDRs

Ticker	Description	% Allocation	Group
SYK	Stryker Corp.	0.47%	Health Care (Medical Prods/Sups)
CAG	Conagra Inc.	0.45%	Foods
AGN	Allergan Inc.	0.44%	Health Care (Drugs/Pharms)
RAL	Ralston Purina Co.	0.44%	Foods
AZA	Alza Corp.	0.44%	Health Care (Drugs)
AVP	Avon Prods Inc.	0.43%	Personal Care
HSY	Hershey Foods Corp.	0.43%	Foods
BGEN	Biogen Inc.	0.43%	Biotechnology
BDX	Becton Dickinson & Co.	0.41%	Health Care (Medical Prods/Sups)
CHIR	Chiron Corp.	0.38%	Biotechnology
ADM	Archer Daniels Midland Co.	0.38%	Agriculture
BSX	Boston Scientific Corp.	0.37%	Health Care (Medical Prods/Sups)
MCK	McKesson HBOC Inc.	0.35%	Distributors (Food & Health)
MEDI	Medimmune Inc.	0.35%	Biotechnology
CLX	Clorox Co.	0.34%	Household Prods (Non-Durable)
CCE	Coca Cola Enterprises Inc.	0.34%	Beverages (Non-Alcoholic)
BMET	Biomet Inc.	0.32%	Health Care (Medical Prods/Sups)
KG	King Pharmaceuticals Inc.	0.31%	Health Care (Drugs)
WPI	Watson Pharmaceuticals Inc.	0.25%	Health Care (Drugs)
FO	Fortune Brands Inc.	0.24%	Housewares
UST	UST Inc.	0.22%	Tobacco
STJ	St Jude Med Inc.	0.21%	Health Care (Medical Prods/Sups)
BF/B	Brown Forman Corp.	0.19%	Beverages — Alcoholic
WIN	Winn Dixie Stores Inc.	0.18%	Retail Stores — Food Chains
BOL	Bausch & Lomb Inc.	0.11%	Health Care (Medical Prods/Sups)
RKY	Coors Adolph Co.	0.11%	Beverages — Alcoholic
BCR	Bard C R Inc.	0.11%	Health Care (Medical Prods/Sups)
ACV	Alberto Culver Co.	0.10%	Personal Care
IFF	International Flavours	0.10%	Chemicals Speciality
SVU	Supervalu Inc.	0.08%	Distributors (Food & Health)
LDG	Longs Drug Stores Corp.	0.05%	Retail Stores — Drug Store
Cyclical/Transportation Select Sector SPDR (XLY)			
WMT	Wal Mart Stores Inc.	20.41%	Retail — Gen Mer Chain
HD	Home Depot Inc.	12.05%	Retail (Building Supplies)
F	Ford Mtr Co. Del	6.36%	Automobiles
TGT	Target Corp.	4.00%	Retail — Gen Mer Chain
GM	General Mtrs Corp.	3.49%	Automobiles
LOW	Lowes Cos Inc.	2.81%	Retail (Building Supplies)
KSS	Kohls Corp.	2.68%	Retail — Dept Stores
GPS	Gap Inc.	2.48%	Retail Speciality — Apparel
COST	Costco Whsl Corp. New	2.22%	Retail — Gen Mer Chain
UNP	Union Pac Corp.	1.84%	Railroads

TABLE 5.3 (CONTINUED)
Companies in the Individual Active Account Sector SPDRs

Ticker	Description	% Allocation	Group
LUV	Southwest Airls Co.	1.83%	Airlines
FDX	Fedex Corp.	1.61%	Air Freight
BNI	Burlington Northn Santa Fe	1.57%	Railroads
S	Sears Roebuck & Co.	1.55%	Retail — Gen Mer Chain
MAS	Masco Corp.	1.42%	Building Materials Group
NKE	Nike Inc.	1.41%	Footwear
MAY	May Dept Stores Co.	1.40%	Retail — Dept Stores
HDI	Harley Davidson Inc.	1.38%	Leisure Time (Products)
TJX	TJX Cos Inc. New	1.26%	Retail Speciality — Apparel
FD	Federated Dept Stores Inc. Del	1.09%	Retail — Dept Stores
MAT	Mattel Inc.	1.09%	Leisure Time (Products)
DPH	Delphi Automotive Sys Corp.	1.07%	Auto Parts & Equipment
BBBY	Bed Bath & Beyond Inc.	1.04%	Retail — Speciality
BBY	Best Buy Co. Inc.	1.00%	Retail (Computers & Electrons)
CSX	CSX Corp.	0.99%	Railroads
DG	Dollar Gen Corp.	0.96%	Retail (Discounters)
NWL	Newell Rubbermaid Inc.	0.96%	Housewares
RSH	Radioshack Corp.	0.91%	Retail (Computers & Electrons)
SPLS	Staples Inc.	0.90%	Retail — Speciality
LTD	Limited Inc.	0.90%	Retail Speciality — Apparel
NSC	Norfolk Southn Corp.	0.88%	Railroads
TOY	Toys R Us Inc.	0.78%	Retail — Speciality
AMR	AMR Corp. Del	0.71%	Airlines
H	Harcourt Gen Inc.	0.68%	Publishing
DAL	Delta Air Lines Inc. De	0.68%	Airlines
GPC	Genuine Parts Co.	0.67%	Auto Parts & Equipment
KM	K Mart Corp.	0.66%	Retail — Gen Mer Chain
SHW	Sherwin Williams Co.	0.65%	Retail (Building Supplies)
VFC	V F Corp.	0.62%	Textiles (Apparel)
JCP	Penney J C Inc.	0.62%	Retail — Dept Stores
RBK	Reebok Intl. Ltd	0.58%	Footwear
GT	Goodyear Tire And Rubber	0.58%	Auto Parts & Equipment
LEG	Leggett & Platt Inc.	0.57%	Household Furnishings & App.
CTX	Centex Corp.	0.53%	Homebuilding
WHR	Whirlpool Corp.	0.49%	Household Furnishings & App.
SWK	Stanley Works	0.48%	Hardware & Tools
TIF	Tiffany & Co. New	0.48%	Retail — Speciality
AZO	Autozone Inc.	0.47%	Retail — Speciality
BDK	Black & Decker Corporation	0.44%	Hardware & Tools
PHM	Pulte Corp.	0.43%	Homebuilding
LIZ	Liz Claiborne Inc.	0.43%	Textiles (Apparel)
U	US Airways Group Inc.	0.42%	Airlines

(continued)

TABLE 5.3 (CONTINUED)
Companies in the Individual Active Account Sector SPDRs

Ticker	Description	% Allocation	Group
ODP	Office Depot Inc.	0.41%	Retail — Speciality
MYG	Maytag Corp.	0.38%	Household Furnishings & App.
TUP	Tupperware Corp.	0.35%	Housewares
DDS	Dillards Inc.	0.35%	Retail — Dept Stores
HAS	Hasbro Inc.	0.35%	Leisure Time (Products)
SNA	Snap On Inc.	0.34%	Auto Parts & Equipment
KBH	KB Home	0.34%	Homebuilding
JWN	Nordstrom Inc.	0.34%	Retail — Dept Stores
BC	Brunswick Corp.	0.31%	Leisure Time (Products)
CC	Circuit City Stores Inc.	0.29%	Retail (Computers & Electrons)
VC	Visteon Corp.	0.24%	Auto Parts & Equipment
R	Ryder Sys Inc.	0.22%	Truckers
CNS	Consolidated Stores Corp.	0.22%	Retail (Discounters)
CTB	Cooper Tire & Rubr Co.	0.19%	Auto Parts & Equipment
Energy Select Sector SPDR (XLE)			
XOM	Exxon Mobil Corp.	23.35%	Oil (Int'l Integrated)
RD	Royal Dutch Pete Co.	14.87%	Oil (Int'l Integrated)
CHV	Chevron Corp.	7.27%	Oil (Int'l Integrated)
TX	Texaco Inc.	4.62%	Oil (Int'l Integrated)
ENE	Enron Corp.	4.54%	Natural Gas — Distr — Pipe Line
EPG	El Paso Corp.	4.41%	Natural Gas — Distr — Pipe Line
SLB	Schlumberger Ltd	4.39%	Oil & Gas (Drilling & Equip)
WMB	Williams Cos Inc.	2.96%	Natural Gas — Distr — Pipe Line
COC/B	Conoco Inc.	2.57%	Oil (Domestic Integrated)
HAL	Halliburton Co.	2.41%	Oil & Gas (Drilling & Equip)
APC	Anadarko Pete Corp.	2.34%	Oil & Gas (Exploration/Prod)
P	Phillips Pete Co.	2.15%	Oil (Domestic Integrated)
RIG	Transocean Sedco Forex Inc.	2.09%	Oil & Gas (Drilling & Equip)
BHI	Baker Hughes Inc.	1.87%	Oil & Gas (Drilling & Equip)
BR	Burlington Res Inc.	1.60%	Oil & Gas (Exploration/Prod)
OXY	Occidental Pete Corp.	1.54%	Oil (Domestic Integrated)
UCL	Unocal Corp.	1.45%	Oil & Gas (Exploration/Prod)
MRO	USX Marathon Group	1.44%	Oil (Domestic Integrated)
NBR	Nabors Industries Inc.	1.34%	Oil & Gas (Drilling & Equip)
DVN	Devon Energy CorporationNew	1.33%	Oil & Gas (Exploration/Prod)
APA	Apache Corp.	1.29%	Oil & Gas (Exploration/Prod)
AHC	Amerada Hess Corp.	1.27%	Oil (Domestic Integrated)
TOS	Tosco Corp.	1.19%	Oil & Gas (Refining & Mktg)
KMI	Kinder Morgan Inc. Kans	1.18%	Natural Gas — Distr — Pipe Line
NE	Noble Drilling Corp.	1.17%	Oil & Gas (Drilling & Equip)
KMG	Kerr Mcgee Corp.	1.17%	Oil & Gas (Exploration/Prod)
EOG	EOG Resources Inc.	1.00%	Oil & Gas (Exploration/Prod)

An Example: Thousand Stock Portfolio

TABLE 5.3 (CONTINUED)
Companies in the Individual Active Account Sector SPDRs

Ticker	Description	% Allocation	Group
SUN	Sunoco Inc.	0.77%	Oil & Gas (Refining & Mktg)
ASH	Ashland Inc.	0.75%	Oil & Gas (Refining & Mktg)
RDC	Rowan Cos Inc.	0.74%	Oil & Gas (Drilling & Equip)
MDR	McDermott Intl. Inc.	0.52%	Engineering & Construction
Financial Select Sector SPDR(XLF)			
C	Citigroup Inc.	12.27%	Financial (Diversified)
AIG	American Intl. Group Inc.	10.18%	Insurance (Multi-Line)
BAC	Bank Amer Corp.	4.84%	Banks (Money Center)
JPM	J P Morgan Chase & Co.	4.64%	Financial (Diversified)
WFC	Wells Fargo & Co. New	4.60%	Banks (Major Regional)
FNM	Federal Natl Mtg Assn	4.36%	Financial (Diversified)
MWD	Morgan Stanley Dean Witter&Co.	3.25%	Financial (Diversified)
AXP	American Express Co.	2.99%	Financial (Diversified)
FRE	Federal Home Ln Mtg Corp.	2.45%	Financial (Diversified)
MER	Merrill Lynch & Co. Inc.	2.43%	Investment Banking/Brokerage
USB	US Bancorp Del	2.43%	Banks (Major Regional)
ONE	Bank One Corp.	2.28%	Banks (Major Regional)
FBF	Fleetboston Finl Corp.	2.22%	Banks (Major Regional)
BK	Bank New York Inc.	1.98%	Banks (Major Regional)
FTU	First Un Corp.	1.76%	Banks (Money Center)
WM	Washington Mut Inc.	1.74%	Savings & Loan Companies
ALL	Allstate Corp.	1.67%	Insurance (Property — Casualty)
KRB	Mbna Corp.	1.53%	Consumer Finance
HI	Household Intl. Inc.	1.52%	Consumer Finance
MMC	Marsh & Mclennan Cos Inc.	1.42%	Insurance Brokers
FITB	Fifth Third Bancorp	1.34%	Banks (Major Regional)
MET	MetLife Inc.	1.25%	Insurance (Life/Health)
SCH	Schwab Charles Corp.	1.16%	Investment Banking/Brokerage
MEL	Mellon Finl Corp.	1.08%	Banks (Major Regional)
PNC	PNC Finl Svcs Group Inc.	1.07%	Banks (Major Regional)
AGC	American Gen Corp.	1.04%	Insurance (Life/Health)
STI	Suntrust Bks Inc.	1.04%	Banks (Major Regional)
NCC	National City Corp.	0.88%	Banks (Major Regional)
LEH	Lehman Brothers Hldgs Inc.	0.85%	Investment Banking/Brokerage
STT	State Street Corporation	0.82%	Financial (Diversified)
AFL	AFLAC Inc.	0.79%	Insurance (Life/Health)
BBT	BB&T Corp.	0.77%	Banks (Major Regional)
PVN	Providian Finl Corp.	0.76%	Consumer Finance
HIG	Hartford Financial Svcs Grp	0.76%	Insurance (Multi-Line)
NTRS	Northern Trust Corp.	0.75%	Banks (Major Regional)
CB	Chubb Corp.	0.69%	Insurance (Property — Casualty)

(continued)

TABLE 5.3 (CONTINUED)
Companies in the Individual Active Account Sector SPDRs

Ticker	Description	% Allocation	Group
WB	Wachovia Corp. New	0.69%	Banks (Major Regional)
SLM	USA Ed Inc.	0.65%	Financial (Diversified)
LTR	Loews Corp.	0.64%	Insurance (Multi-Line)
KEY	KeyCorp. New	0.60%	Banks (Major Regional)
COF	Capital One Finl Corp.	0.60%	Consumer Finance
CMA	Comerica Inc.	0.59%	Banks (Major Regional)
GDW	Golden West Finl Corp. Del	0.56%	Savings & Loan Companies
SPC	St Paul Cos Inc.	0.52%	Insurance (Property — Casualty)
BEN	Franklin Res Inc.	0.52%	Investment Management
AOC	Aon Corp.	0.50%	Insurance Brokers
LNC	Lincoln Natl Corp. In	0.44%	Insurance (Life/Health)
MBI	MBIA Inc.	0.43%	Insurance (Property — Casualty)
SOTR	Southtrust Corp.	0.42%	Banks (Major Regional)
SNV	Synovus Finl Corp.	0.42%	Banks (Major Regional)
CIT	CIT Group Inc.	0.41%	Financial (Diversified)
MTG	MGIC Invt Corp. Wis	0.40%	Insurance (Property — Casualty)
PGR	Progressive Corp. Ohio	0.39%	Insurance (Property — Casualty)
UNM	Unumprovident Corp.	0.38%	Insurance (Life/Health)
JP	Jefferson Pilot Corp.	0.38%	Insurance (Life/Health)
RGBK	Regions Finl Corp.	0.37%	Banks (Major Regional)
ABK	Ambac Finl Group Inc.	0.36%	Financial (Diversified)
ASO	Amsouth Bancorporation	0.34%	Banks (Major Regional)
CINF	Cincinnati Finl Corp.	0.33%	Insurance (Property — Casualty)
SV	Stilwell Financial Inc.	0.32%	Investment Management
CF	Charter One Finl Inc.	0.32%	Savings & Loan Companies
CCR	Countrywide Cr Inds Inc.	0.31%	Consumer Finance
OK	Old Kent Finl Corp.	0.29%	Banks (Major Regional)
CNC	Conseco Inc.	0.29%	Insurance (Life/Health)
UPC	Union Planters Corp.	0.28%	Banks (Major Regional)
BSC	Bear Stearns Cos Inc.	0.27%	Investment Banking/Brokerage
TMK	Torchmark Inc.	0.26%	Insurance (Life/Health)
MCO	Moodys Corp.	0.24%	Financial (Diversified)
TROW	Price T Rowe Group Inc.	0.21%	Investment Management
SAFC	Safeco Corp.	0.20%	Insurance (Property - Casualty)
HBAN	Huntington Bancshares Inc.	0.20%	Banks (Major Regional)
Industrial Select Sector SPDR (XLI)			
GE	General Elec Co.	17.29%	Electrical Equipment
TYC	Tyco Intl. Ltd New	9.13%	Manufacturing (Diversified)
BA	Boeing Co.	4.98%	Aerospace/Defense
MMM	Minnesota Mng & Mfg Co.	4.95%	Manufacturing (Diversified)
HON	Honeywell Intl. Inc.	4.50%	Manufacturing (Diversified)
EMR	Emerson Elec Co.	4.39%	Electrical Equipment

TABLE 5.3 (CONTINUED)
Companies in the Individual Active Account Sector SPDRs

Ticker	Description	% Allocation	Group
WMI	Waste Mgmt Inc. Del	3.77%	Waste Management
UTX	United Technologies Corp.	3.69%	Manufacturing (Diversified)
AW	Allied Waste Industries Inc.	3.42%	Waste Management
ITW	Illinois Tool Wks Inc.	3.02%	Manufacturing (Diversified)
CAT	Caterpillar Inc.	2.95%	Machinery — Diversified
GLW	Corning Inc.	2.43%	Communications Equipment
DHR	Danaher Corp.	2.28%	Manufacturing (Diversified)
DE	Deere & Co.	2.08%	Machinery — Diversified
TXT	Textron Inc.	1.97%	Manufacturing (Diversified)
JCI	Johnson Ctls Inc.	1.91%	Manufacturing (Diversified)
IR	Ingersoll Rand Co.	1.79%	Machinery — Diversified
PH	Parker Hannifin Corp.	1.70%	Manufacturing (Diversified)
DOV	Dover Corp.	1.69%	Machinery — Diversified
LMT	Lockheed Martin Corp.	1.64%	Aerospace/Defense
CBE	Cooper Inds Inc.	1.52%	Electrical Equipment
ETN	Eaton Corp.	1.51%	Manufacturing (Diversified)
PCAR	Paccar Inc.	1.48%	Trucks & Parts
NSI	National Svc Inds Inc.	1.43%	Manufacturing (Diversified)
PLL	Pall Corp.	1.40%	Manufacturing (Specialized)
CR	Crane Co.	1.39%	Manufacturing (Diversified)
CUM	Cummins Engine Inc.	1.35%	Trucks & Parts
GD	General Dynamics Corp.	1.33%	Aerospace/Defense
BGG	Briggs & Stratton Corp.	1.17%	Manufacturing (Specialized)
RTN/B	Raytheon Co.	1.07%	Electronics (Defense)
DCN	Dana Corp.	0.95%	Auto Parts & Equipment
NAV	Navistar Intl. Corp. Inc.	0.78%	Trucks & Parts
ROK	Rockwell Intl. Corp. New	0.71%	Electrical Equipment
NOC	Northrop Grumman Corp.	0.67%	Aerospace/Defense
ABI	Applera Corp. Applied Biosys	0.63%	Health Care (Medical Prods/Sups)
TRW	TRW Inc.	0.45%	Auto Parts & Equipment
TMO	Thermo Electron Corp.	0.43%	Manufacturing (Diversified)
GR	Goodrich B F Co.	0.42%	Aerospace/Defense
ITT	ITT Inds Inc.	0.36%	Manufacturing (Diversified)
FLR	Fluor Corp. New	0.35%	Engineering & Construction
PKI	Perkinelmer Inc.	0.28%	Electronics (Instrument.)
APCC	American Pwr Conversion Corp.	0.27%	Electrical Equipment
MIL	Millipore Corp.	0.23%	Manufacturing (Specialized)
PWER	Power One Inc.	0.12%	Electrical Equipment
Technology Select Sector SPDR (XLK)			
MSFT	Microsoft Corp.	12.36%	Computers Software/Services
INTC	Intel Corp.	7.50%	Electronics — Semiconductors

(continued)

TABLE 5.3 (CONTINUED)
Companies in the Individual Active Account Sector SPDRs

Ticker	Description	% Allocation	Group
AOL	AOL Time Warner Inc.	7.39%	Entertainment
IBM	International Business Machs	7.15%	Computers (Hardware)
CSCO	Cisco Sys Inc.	4.87%	Computers (Networking)
ORCL	Oracle Corp.	3.58%	Computers Software/Services
T	AT & T Corp.	3.39%	Telephone Long Distance
DELL	Dell Computer Corp.	2.83%	Computers (Hardware)
EMC	E M C Corp. Mass	2.75%	Computers (Peripherals)
HWP	Hewlett Packard Co.	2.57%	Computers (Hardware)
Q	Qwest Communications Intl. Inc.	2.46%	Telephone
TXN	Texas Instrs Inc.	2.30%	Electronics — Semiconductors
SUNW	Sun Microsystems Inc.	2.14%	Computers (Hardware)
NT	Nortel Networks Corp.	1.91%	Communications Equipment
QCOM	Qualcomm Inc.	1.81%	Communications Equipment
WCOM	Worldcom Inc.	1.79%	Telephone Long Distance
AMAT	Applied Materials Inc.	1.52%	Equipment (Semiconductors)
ADP	Automatic Data Processing Inc.	1.49%	Services (Data Processing)
LU	Lucent Technologies Inc.	1.44%	Communications Equipment
MOT	Motorola Inc.	1.33%	Communications Equipment
CPQ	Compaq Computer Corp.	1.32%	Computers (Hardware)
EDS	Electronic Data Sys Corp. New	1.13%	Services (Computer Systems)
MU	Micron Technology Inc.	1.10%	Electronics — Semiconductors
FDC	First Data Corp.	1.02%	Services (Data Processing)
JDSU	JDS Uniphase Corp.	1.02%	Communications Equipment
FON	Sprint Corp.	0.83%	Telephone Long Distance
VRTS	Veritas Software Corp.	0.80%	Computers Software/Services
PCS	Sprint Corp.	0.77%	Cellular/Wireless Telecomms
TLAB	Tellabs Inc.	0.72%	Communications Equipment
CA	Computer Assoc Intl. . Inc.	0.68%	Computers Software/Services
PAYX	Paychex Inc.	0.62%	Services (Data Processing)
A	Agilent Technologies Inc.	0.59%	Electronics (Instrument.)
ADI	Analog Devices Inc.	0.55%	Electronics — Semiconductors
LLTC	Linear Technology Corp.	0.55%	Electronics — Semiconductors
SLR	Solectron Corp.	0.54%	Electrical Equipment
EK	Eastman Kodak Co.	0.52%	Photography/Imaging
GX	Global Crossing Ltd	0.51%	Telephone Long Distance
MXIM	Maxim Integrated Prods Inc.	0.50%	Electronics — Semiconductors
SEBL	Siebel Sys Inc.	0.49%	Computers Software/Services
XLNX	Xilinx Inc.	0.49%	Electronics — Semiconductors
NXTL	Nextel Communications Inc.	0.48%	Cellular/Wireless Telecomms
AMD	Advanced Micro Devices Inc.	0.42%	Electronics — Semiconductors
CMVT	Comverse Technology Inc.	0.41%	Communications Equipment
ADBE	Adobe Sys Inc.	0.39%	Computers Software/Services
PBI	Pitney Bowes Inc.	0.38%	Office Equipment & Supplies

An Example: Thousand Stock Portfolio

TABLE 5.3 (CONTINUED)
Companies in the Individual Active Account Sector SPDRs

Ticker	Description	% Allocation	Group
YHOO	Yahoo Inc.	0.37%	Computers Software/Services
CEFT	Concord Efs Inc.	0.36%	Services (Data Processing)
ALTR	Altera Corp.	0.36%	Electronics — Semiconductors
AAPL	Apple Computer	0.35%	Computers (Hardware)
SFA	Scientific Atlanta Inc.	0.34%	Communications Equipment
KLAC	KLA Tencor Corp.	0.34%	Equipment (Semiconductors)
PSFT	Peoplesoft Inc.	0.32%	Computers Software/Services
BRCM	Broadcom Corp.	0.30%	Electronics — Semiconductors
MOLX	Molex Inc.	0.29%	Electrical Equipment
ADCT	ADC Telecommunications Inc.	0.28%	Communications Equipment
NTAP	Network Appliance Inc.	0.26%	Computers (Networking)
TSG	Sabre Hldgs Corp.	0.26%	Services (Computer Systems)
SANM	Sanmina Corp.	0.25%	Electrical Equipment
LXK	Lexmark Intl. Inc.	0.25%	Computers (Peripherals)
TER	Teradyne Inc.	0.24%	Equipment (Semiconductors)
INTU	Intuit	0.24%	Computers Software/Services
CSC	Computer Sciences Corp.	0.24%	Services (Computer Systems)
GTW	Gateway Inc.	0.24%	Computers (Hardware)
NVLS	Novellus Sys Inc.	0.24%	Equipment (Semiconductors)
CVG	Convergys Corp.	0.24%	Services (Commercial & Consum)
BMC	BMC Software Inc.	0.23%	Computers Software/Services
LSI	LSI Logic Corp.	0.23%	Electronics — Semiconductors
FISV	Fiserv Inc.	0.23%	Services (Data Processing)
NSM	National Semiconductor Corp.	0.22%	Electronics — Semiconductors
SBL	Symbol Technologies Inc.	0.22%	Electrical Equipment
AMCC	Applied Micro Circuits Corp.	0.21%	Electronics — Semiconductors
EFX	Equifax Inc.	0.21%	Services (Data Processing)
PALM	Palm Inc.	0.20%	Computers (Hardware)
UIS	Unisys Corp.	0.20%	Computers Software/Services
VTSS	Vitesse Semiconductor Corp.	0.19%	Electronics — Semiconductors
CTL	Centurytel Inc.	0.19%	Telephone
JBL	Jabil Circuit Inc.	0.17%	Manufacturing (Specialized)
XRX	Xerox Corp.	0.17%	Photography/Imaging
CTXS	Citrix Sys Inc.	0.17%	Computers Software/Services
CPWR	Compuware Corp.	0.16%	Computers Software/Services
NCR	NCR Corp. New	0.16%	Computers (Hardware)
AV	Avaya Inc.	0.16%	Computers (Networking)
TEK	Tektronix Inc.	0.15%	Electronics (Instrument.)
GWW	Grainger W W Inc.	0.15%	Electronics (Component Dist)
MERQ	Mercury Interactive Corp.	0.14%	Computers Software/Services
CZN	Citizens Communications Co.	0.14%	Telephone
CS	Cabletron Systems Inc.	0.12%	Computers (Networking)

(*continued*)

TABLE 5.3 (CONTINUED)
Companies in the Individual Active Account Sector SPDRs

Ticker	Description	% Allocation	Group
PMTC	Parametric Technology Corp.	0.12%	Computers Software/Services
ADSK	Autodesk Inc.orporated	0.10%	Computers Software/Services
CNXT	Conexant Sys Inc.	0.09%	Electronics — Semiconductors
QLGC	QLogic Corp.	0.09%	Electronics — Semiconductors
NOVL	Novell Inc.	0.07%	Computers Software/Services
ANDW	Andrew Corp.	0.07%	Communications Equipment
BVSN	Broadvision Inc.	0.06%	Computers Software/Services
TNB	Thomas & Betts Corp.	0.05%	Electrical Equipment
SAPE	Sapient Corp.	0.04%	Services (Computer Systems)
ADPT	Adaptec Inc.	0.04%	Electronics — Semiconductors
Utilities Select Sector SPDR (XLU)			
SBC	SBC Communications Inc.	18.84%	Telephone
VZ	Verizon Communications	16.61%	Telephone
DUK	Duke Energy Co.	4.71%	Electric Companies
AES	AES Corp.	3.93%	Power Producers (Independ.)
EXC	Exelon Corp.	3.69%	Electric Companies
SO	Southern Co.	3.53%	Electric Companies
BLS	Bellsouth Corp.	3.32%	Telephone
AEP	American Elec Pwr Inc.	2.83%	Electric Companies
D	Dominion Res Inc. Va New	2.54%	Electric Companies
XEL	XCEL Energy Inc.	2.52%	Electric Companies
REI	Reliant Energy Inc.	2.39%	Electric Companies
AT	Alltel Corp.	2.35%	Telephone
DYN	Dynegy Inc. New	2.06%	Natural Gas — Distr — Pipe Line
CPN	Calpine Corp.	1.94%	Power Producers (Independ.)
FPL	FPL Group Inc.	1.82%	Electric Companies
TXU	TXU Corp.	1.80%	Electric Companies
PGN	Progress Energy Inc.	1.72%	Electric Companies
PEG	Public Svc Enterprise Group	1.68%	Electric Companies
ETR	Entergy Corp.	1.67%	Electric Companies
PPL	PPL Corp.	1.58%	Electric Companies
CEG	Constellation Energy Group Inc.	1.46%	Electric Companies
ED	Consolidated Edison Inc.	1.42%	Electric Companies
FE	Firstenergy Corp.	1.27%	Electric Companies
CIN	Cinergy Corp.	1.24%	Electric Companies
AEE	Ameren Corp.	1.18%	Electric Companies
DTE	DTE Energy Co.	1.17%	Electric Companies
CMS	CMS Energy Corp.	1.11%	Electric Companies
PNW	Pinnacle West Cap Corp.	1.09%	Electric Companies
SRE	Sempra Energy	1.08%	Natural Gas — Distr — Pipe Line
GPU	GPU Inc.	0.92%	Electric Companies
PCG	PG&E Corp.	0.81%	Electric Companies

TABLE 5.3 (CONTINUED)
Companies in the Individual Active Account Sector SPDRs

Ticker	Description	% Allocation	Group
NI	NiSource Inc.	0.79%	Natural Gas — Distr — Pipe Line
EIX	Edison Intl.	0.77%	Electric Companies
NMK	Niagara Mohawk Hldgs Inc.	0.73%	Electric Companies
OKE	Oneok Inc. New	0.72%	Natural Gas — Distr — Pipe Line
PGL	Peoples Energy Corp.	0.68%	Natural Gas — Distr — Pipe Line
KSE	Keyspan Corp.	0.64%	Natural Gas — Distr — Pipe Line
AYE	Allegheny Energy Inc.	0.64%	Electric Companies
GAS	Nicor Inc.	0.61%	Natural Gas — Distr — Pipe Line

ALLOCATING ASSETS

The two BCP accounts, the passive and active, are each allocated a fixed percentage of assets depending on investor risk parameters and expected market conditions as illustrated in Table 5.4.

TABLE 5.4
Portfolio Allocations

Basket Case Portfolio — Two Accounts

    I. Passive Account
        Diversified Core Account (50–75%)
    II. Active Account
        Select SPDR Account (50–25%)

FIVE-STAGE SYSTEM

The BCP is a low-maintenance, five-stage system using three investment tools, or indicators. Upkeep only requires about 30 minutes per week. This maintenance allows a clear determination of the portfolio's position in relation to the benchmark, whether it is necessary to shift to or from active management, what the index blend should be, and what the components of the select SPDR account should be.

The five stages are:

1. Determine the passive/active allocation.
2. Establish the index blend within the Diversified Core Account.
3. Determine approximate position in the business cycle.
4. Choose Select Sector SPDRs for Active Account.
5. Monitor and adjust portfolio.

Stage one is typically reviewed on a quarterly basis.

INVESTMENT TOOLS

Four basic tools are used to create, establish, and maintain the portfolio:

1. Active/passive indicator
2. Value at risk
3. Business cycle position
4. Relative strength

Each step and related investment tool is explained in the following chapters. Instructions for interpreting the active/passive indicator is supplied in Appendix C.

6 The Diversified Core

As we have seen, basket securities are well designed to be used not only as a passive index-like portfolio, but also in the context of a blend of passive and active components.

This chapter explores in a preliminary way the issue of what proportions of the equity portfolio should be invested in the passive and active components respectively. Two cautionary thoughts need to be expressed. First is the reiteration that using an active component at all in the portfolio presumes the investor has some basis for making active selections. Secondly, the matter of how much should be allocated to passive and active is complex, and deserves further research.

In the following pages, the authors present an approach using several tools that they believe can provide some insight with regard to this issue. However, they readily acknowledge that it is less than a definitive answer. At the heart of the matter is the investor's ability to generate positive "alpha," that is, positive risk-adjusted excess returns.

In the following discussions we use the concepts of a diversified core for the passive portion and an active sector for the active component.

Finally, the chapter describes an iterative process by which the investor can assess the risk of various mixes of the index baskets that constitute the diversified core, using the concept of value at risk (VAR).

It should be noted that the diversified core could be constructed using other basket securities. For example, it could include iShares Small Cap Index and/or iShares Value or Growth Indexes, or mixing segments of large cap, mid-cap, or small cap baskets. The blend decision can be addressed by the particular demands of the investment portfolio. In this example, only the S&P 500 (SPY), S&P 400 (MDY), and Nasdaq 100 (QQQ) are used.

The goal of the diversified core portion of the equity portfolio is to invest in the optimum risk/reward blend of three market indices, shown in Table 6.1: the S&P 500, the S&P 400, and the Nasdaq 100. For example, a range of about 50 to 75 percent of portfolio assets are earmarked for investment in the diversified core as determined by the active/passive indicator. The actual blend of the three indices is established by calculating a statistical measure of possible portfolio losses known as value at risk, or VAR.

The diversified core also serves a secondary function of equitizing cash. The basket securities within the diversified core constitute a "sweep" account for funds that are not invested in the active sector of the portfolio. This allows market participation without attempting to time the market. While market timing is a dynamic lever available to the active manager, it is less consequential than the strategic asset allocation decision.

TABLE 6.1
The Diversified Core

Fund Name	Ticker	Representative Index
SPDR Trust Series 1	SPY	S&P 500 Index
MidCap SPDR Trust Series 1	MDY	S&P 400 Index
Nasdaq-100 Trust Series 1	QQQ	Nasdaq 100 Index

Additionally, equitization of cash in the diversified core allows an effective blending of active and passive strategies, moving between higher concentrations in passive or active investments as specified by the active/passive indicator. Allocation can be varied by assessing the environment for active versus passive investing as suggested by the active/passive indicator. If the active/passive indicator implies that passive investing is more likely to outperform, allocation to the diversified core in this example can go as high as 75 percent of the total equity portfolio. On the other hand, it can fall to as low as 50 percent if the active/passive indicator suggests active management is likely to be stronger.

The active versus passive debate has taken on increased importance during the last few years as the volatility of individual stocks has dramatically increased. For instance, during the year 2000, according to *The Wall Street Journal*, more than 460 of the largest 1,000 publicly traded companies experienced one-day price drops in excess of 20 percent! Such volatility places greater strain on the security selection processes of the active manager.

Active managers argue that ability and hard work enable them to outperform an asset class to such a degree that investors are more than compensated for all costs and risks associated with active management. Advocates of passive management assert that investors should invest passively within each asset class and focus their resources to high-level asset allocation decisions.

Both factions present persuasive arguments, offer strengths and weaknesses, and have large numbers of devoted supporters. Is one strategy actually better than the other? The answer is, "that depends."

Active managers go through cycles of over- and underperformance. Figure 6.1 shows the annual rank of the S&P 500 index 36-month rolling returns versus returns of actively managed Large-Cap Blend mutual funds from 1992 through 2000.

From March 1992 to September 1994, (letter A) active managers generally outperformed the S&P 500. This trend was reversed up to June 1995 (B) when a majority of managers underperformed as indicated by the percentile rank of the S&P 500 rising above 50 percent. For a brief period from June to December 1995 (C), active management rebounded. December 1995 through September 1997 (D) was another period of passive supremacy, which was briefly reversed through March 1999 (E), when active management once again outperformed. From March 1999 to the end of the graph at March 2000, active management prevailed.

Therefore, in response to the earlier question, "Is passive management superior to active?" it makes sense to underweight active management during periods of

The Diversified Core

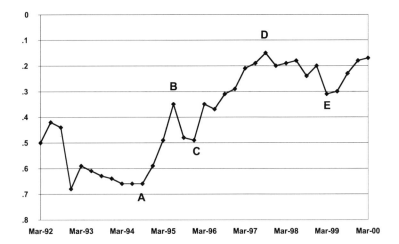

FIGURE 6.1 S&P 500 Ranking versus Large-Cap Blend Fund Managers

underperformance, and overweight during periods of outperformance. Of course the logical follow-up question is how does one predict, or at least monitor, this environment, as opposed to simply measuring historical patterns?

One approach to this question is to recognize that an important reason for the cyclic nature of manager performance is found within the very structure of the S&P 500 Index itself. The Basket Case Portfolio active/passive indicator exploits this characteristic in determining the allocation to the diversified core.

The S&P 500 Index is a capitalization-weighted index; meaning larger companies in the index have a much greater influence on overall index returns than smaller companies. This index characteristic provides a key to measure the position in the management performance cycle. Active portfolio managers typically establish more equally weighted portfolios than their cap-weighted benchmarks.

During periods when larger-cap companies outperform, active managers have a difficult time beating the benchmark. But, during periods when smaller-cap companies are outperforming, active managers generally beat the benchmark.

The active/passive allocation is strongly influenced by cyclical changes in the market environment. Studies have shown that when the difference between the rolling 36-month annualized returns of the cap-weighted and equal-weighted S&P 500 is graphed (Figure 6.2), the resultant plot closely matches the S&P 500 Index over- and underperformance periods shown in Figure 6.1.

The cap-weighted equal-weighted index comparison makes a very effective active/passive indicator. While creating historical equal-weighted and capitalization-weighted S&P 500 portfolios to review past data is a time-consuming project, it is not difficult to construct current portfolios to go forward. Bloomberg, the premier market information and analytical resource, has the ability to establish both equal and cap-weighted indexes through its "Model Portfolio" function. Use the following steps:

- Pull up the S&P 500 Index on the Bloomberg
- Then type IPRT <GO> to create a portfolio.

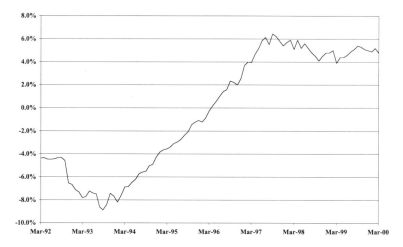

FIGURE 6.2 Relative Performance of Rolling 36-Month Cap-Weighted vs. Equal-Weighted S&P 500

- Then use PCPY < GO> to copy this portfolio.
- PBAL < GO> to balance this as a dollar weighted index

Repeat the entire procedure for the capitalization-weighted S&P portfolio with the exception that under the PBAL step, input the current S&P portfolio weights. Maintain these portfolios, changing when the index changes and linking the monthly returns.

STEP ONE — DETERMINING THE ALLOCATION TO THE DIVERSIFIED CORE

THE BCP ACTIVE/PASSIVE INDICATOR

For the purposes of this book, a simpler yet useful active/passive indicator is used. This proxy for the cap-weighted, equal-weighted comparison is a point-and-figure plot measuring the relative strength between the cap-weighted S&P 500 Index and a 50/50 blend of the S&P 500 (SPY) and the S&P 400 (MDY) indexes.

The average market capitalization of the combination is approximately 60 percent that of the S&P 500 alone. This figure closely mirrors the relative market cap of the typical large-cap managed portfolio. As such, it is a useful proxy for the equal-weighted S&P 500 Index. All the data is easily obtained and the computations are extremely simple.

Relative strength (RS), measures the relationship, or ratio, between the price movement of a security and that of another security or market average. The BCP active/passive indicator analyzes whether active or passive management is outperforming by plotting the SPY/MDY blend performance relative to the S&P 500 on a weekly basis.

The Diversified Core

Suppose for example that the blended index is trading at 105 per share and the S&P 500 is at 1000. To determine how the blended index is trading relative to the S&P 500, simply divide the price of the blended index by the S&P 500 and then plot the resulting number on a point and figure chart. Mulitply the result by 100 to create an easier number.

$$\text{Relative Strength} = \frac{\text{Price of Blend}}{\text{Price of market average}}$$

$$\text{Relative Strength} = \frac{105}{1,000} = 0.105 \times 100 = 10.50$$

Computer spreadsheet programs dramatically speed the process of calculating relative strength. For example, to construct one in Excel, label six columns: Date, S&P 500 close, Day of Week, Blend Close, Weekly Close, and Relative Strength (Rel. Str.). In the Day of Week and Relative Strength columns, input the formulas as shown. These will pull out the weekly close and then calculate the relative strength.

Table 6.2 is pulling out the value for each Friday. Other days can be selected by changing the formula value in Column C. In the date function for Excel, Sunday is day number 1, Monday is 2, and Saturday is 7. For Thursday, the formula would read, =IF(C2=5, etc.

To make the relative strength number easier to read and chart, move the decimal over two or three places to the right (multiply the result by either 100 or 1,000).

Figure 6.3 shows a completed point and figure relative strength plot of the active/passive indicator from March 1992 through December 2000.

TABLE 6.2
Relative Strength Spreadsheet Setup

	A	B	C	D	E
1	Date	S&P 500	Day of Week	Blend Close	Rel. Str.
2	06/01/95	534.21	=WEEKDAY(C4,1)	46.09	=IF(C4=6,((D4/B4)*100),"")
3	06/02/95	536.91	6	46.14	8.59
4	06/05/95	537.73	2	46.43	
5	06/06/95	537.09	3	46.45	
6	06/07/95	535.55	4	46.20	
7	06/08/95	533.56	5	46.20	
8	06/09/95	532.35	6	46.01	8.64
9	06/12/95	532.54	2	46.15	
10	06/13/95	536.23	3	46.49	
11	06/14/95	536.48	4	46.50	
12	06/15/95	539.07	5	46.74	
13	06/16/95	539.98	6	46.56	8.62
14	06/19/95	545.22	2	47.11	

8.90				
8.80	X			5
8.70	8	O		3
8.60	6	O		2
8.50		12		12
8.40		11		X
8.30		12		9
8.20		4	X	8
8.10		6	3	O	5
8.00		7	2	O	X
7.90		8	6	O	X
7.80		10	X	4
7.70		3	X	
7.60		O		
7.58				
	9	9	0	0
	5	9	0	1

FIGURE 6.3 The Active/Passive Indicator

Columns of X's represent periods when the SPY/MDY blend relative strength is rising while a column of O's indicates the relative strength is falling. The chart begins in March 1995, when the MidCap SPDR was introduced. The chart conforms closely with the results in both Figures 6.1 and 6.2 from the same time period. Note that in September 2000, active management overtook passive management and has continued this trend up to the present.

The following guidelines are derived from the foregoing description: When the active/passive indicator shows the cap-weighted S&P 500 is outperforming the blended SPY/MDY index, increase passive management by boosting the allocation of the diversified core toward 75 percent. When the blended index is outperforming, reduce the passive allocation in the diversified core toward 50percent .

The active/passive indicator uses a point and figure relative strength chart because of its ease of construction, clarity of interpretation and ability to filter out a great deal of market noise. A rising column of X's represents increasing relative strength, while a declining column of O's represents falling relative strength. A complete set of instructions for creating and interpreting the active/passive indicator using point-and-figure charts is supplied in Appendix C.

STEP TWO — DETERMINING THE DIVERSIFIED CORE BLEND

Once the active/passive allocation is set the next task is to establish the relative blend of the three diversified core indices. This can be accomplished using a value at risk (VAR) calculation. Value at risk seeks to answer the question, "How much can I lose?" by showing the dollar amount of maximum potential loss over a specified holding period and confidence interval.

The VAR methodology can be used to aggregate and estimate risk consistently over a broad set of asset classes. Therefore, the VAR methodology can provide a common framework to aggregate risk, supplying meaningful information regarding

The Diversified Core

the risk characteristics of the diversified core blend. VAR is rapidly becoming a popular way to set risk probability parameters or limits on a portfolio. VAR for the BCP blend can be constructed using either absolute VAR, the total variation of the portfolio, or relative VAR versus a benchmark.

There are three basic methods to compute VAR: Variance/Covariance, Historical, and Monte Carlo Simulation. The BCP uses the historical method to adjust the individual index weights according to the model outcomes. Since the diversified core contains only three assets, SPY, MDY, and QQQ, the calculations are straightforward.

Historical VAR is created using chronological price data. The process begins by assigning percentage weights to the three diversified core indices. For example, the beginning allocation could be 65 percent SPY, 30 percent MDY, and 5 percent QQQ. The market value of the diversified core portfolio with these weightings is then calculated for each of the last 100 trading days. Portfolio returns are then ranked from lowest to highest. The amount of the loss on the fifth worst trading represents the VAR at the 95 percent confidence level.

The same process is applied to the benchmark, which in this case is the S&P 500. S&P 500 returns for the last 100 days are tabulated, ranked from lowest to highest, and compared with those of the diversified core. If the diversified core VAR figure is too high relative to the benchmark, adjust the constituent index weights and recalculate. Once an acceptable risk level, absolute or relative, is obtained, the blend is applied to the diversified core.

Although simple, there are some drawbacks to this method. The most obvious is that the prior 100 trading days may not be representative of future market movements. VAR may be over- or underestimated depending upon market history.

Let's look at an example. Suppose the BCP is to invest $133,333. The active/passive indicator has shown that passive investing is currently stronger than active and therefore the weighting to the diversified core portfolio has been set to 75 percent, or $100,000. We begin by arbitrarily weighting 65 percent SPY, 30 percent MDY, and 5 percent QQQ. Table 6.3 shows the daily returns for each of the three core indices and the resultant portfolio gain or loss for each of the last 100 trading days.

TABLE 6.3
Developing Diversified Core Value at Risk

Date	SPY % G/L	SPY Wgt G/L	MDY % G/L	MDY Wgt G/L	QQQ % G/L	QQQ Wgt G/L	Gain/Loss (%)	Gain/Loss ($)
12/05/00	3.28%	1.97%	3.27%	1.14%	10.99%	0.55%	3.66%	3,661.31
12/06/00	-1.60%	-0.96%	-0.34%	-0.12%	-3.80%	-0.19%	-1.27%	(1,267.73)
12/07/00	-1.37%	-0.82%	0.14%	0.05%	-2.21%	-0.11%	-0.89%	(886.24)
12/08/00	0.23%	0.14%	2.90%	1.01%	2.26%	0.11%	1.27%	1,267.90
12/11/00	3.48%	2.09%	2.88%	1.01%	9.38%	0.47%	3.56%	3,563.28
12/12/00	-0.43%	-0.26%	-2.80%	-0.98%	-3.66%	-0.18%	-1.42%	(1,420.63)

(continued)

TABLE 6.3 (CONTINUED)
Developing Diversified Core Value at Risk

	SPY		MDY		QQQ		Gain/Loss	Gain/Loss
Date	% G/L	Wgt G/L	% G/L	Wgt G/L	% G/L	Wgt G/L	(%)	($)
12/13/00	-1.37%	-0.82%	-1.90%	-0.67%	-4.08%	-0.20%	-1.69%	(1,690.88)
12/14/00	-1.27%	-0.76%	-1.67%	-0.58%	-4.89%	-0.24%	-1.59%	(1,593.22)
12/15/00	-2.56%	-1.53%	-1.36%	-0.48%	-2.29%	-0.11%	-2.12%	(2,124.79)
12/18/00	1.34%	0.80%	1.79%	0.63%	0.20%	0.01%	1.44%	1,438.22
12/19/00	-2.04%	-1.22%	-1.49%	-0.52%	-7.62%	-0.38%	-2.12%	(2,123.91)
12/20/00	-2.90%	-1.74%	-3.09%	-1.08%	-5.73%	-0.29%	-3.11%	(3,106.22)
12/21/00	0.69%	0.42%	0.71%	0.25%	0.59%	0.03%	0.69%	693.33
12/22/00	3.00%	1.80%	3.98%	1.39%	7.92%	0.40%	3.59%	3,586.82
12/26/00	1.07%	0.64%	0.91%	0.32%	0.62%	0.03%	1.00%	995.18
12/27/00	0.73%	0.44%	1.81%	0.63%	1.13%	0.06%	1.13%	1,129.45
12/28/00	0.30%	0.18%	1.65%	0.58%	-0.20%	-0.01%	0.75%	749.13
12/29/00	-1.89%	-1.14%	-2.14%	-0.75%	-4.98%	-0.25%	-2.13%	(2,133.58)
01/02/01	-1.81%	-1.09%	-3.84%	-1.34%	-8.46%	-0.42%	-2.85%	(2,853.51)
01/03/01	4.80%	2.88%	3.98%	1.39%	16.84%	0.84%	5.12%	5,116.26
01/04/01	-1.08%	-0.65%	-2.14%	-0.75%	-1.80%	-0.09%	-1.48%	(1,483.54)
01/05/01	-3.26%	-1.96%	-2.74%	-0.96%	-7.65%	-0.38%	-3.30%	(3,300.27)
01/08/01	0.77%	0.46%	-0.07%	-0.02%	1.10%	0.06%	0.50%	495.27
01/09/01	-0.26%	-0.16%	-0.45%	-0.16%	0.00%	0.00%	-0.32%	(316.81)
01/10/01	1.76%	1.05%	2.99%	1.05%	5.13%	0.26%	2.36%	2,357.34
01/11/01	0.09%	0.06%	0.73%	0.26%	2.91%	0.15%	0.46%	457.71
01/12/01	-0.19%	-0.11%	0.54%	0.19%	1.29%	0.06%	0.14%	139.72
01/16/01	0.64%	0.38%	-0.20%	-0.07%	-1.20%	-0.06%	0.25%	253.35
01/17/01	0.46%	0.28%	0.47%	0.16%	3.05%	0.15%	0.59%	592.35
01/18/01	1.00%	0.60%	-0.20%	-0.07%	4.21%	0.21%	0.74%	737.29
01/19/01	-0.57%	-0.34%	-0.80%	-0.28%	-0.38%	-0.02%	-0.64%	(641.14)
01/22/01	0.66%	0.40%	0.79%	0.28%	0.09%	0.00%	0.68%	681.37
01/23/01	0.79%	0.47%	2.36%	0.83%	2.10%	0.10%	1.40%	1,404.34
01/24/01	0.30%	0.18%	0.07%	0.02%	-0.95%	-0.05%	0.15%	154.90
01/25/01	-0.25%	-0.15%	-0.59%	-0.21%	-3.91%	-0.20%	-0.55%	(552.90)
01/26/01	-0.11%	-0.07%	0.18%	0.06%	1.62%	0.08%	0.08%	75.59
01/29/01	0.53%	0.32%	2.20%	0.77%	2.22%	0.11%	1.20%	1,200.33
01/30/01	0.88%	0.53%	-0.05%	-0.02%	-0.37%	-0.02%	0.49%	490.43
01/31/01	-0.57%	-0.34%	-0.41%	-0.14%	-3.67%	-0.18%	-0.67%	(667.25)
02/01/01	0.66%	0.40%	-0.36%	-0.13%	1.32%	0.07%	0.34%	337.96
02/02/01	-2.27%	-1.36%	-1.56%	-0.54%	-5.53%	-0.28%	-2.18%	(2,182.45)
02/05/01	0.73%	0.44%	0.61%	0.21%	-0.08%	0.00%	0.65%	650.50
02/06/01	-0.29%	-0.18%	1.02%	0.36%	0.16%	0.01%	0.19%	186.96
02/07/01	-0.52%	-0.31%	-0.47%	-0.16%	-1.62%	-0.08%	-0.55%	(554.68)
02/08/01	-1.17%	-0.70%	-0.31%	-0.11%	-2.99%	-0.15%	-0.96%	(958.10)
02/09/01	-0.96%	-0.58%	-0.43%	-0.15%	-4.07%	-0.20%	-0.93%	(930.14)
02/12/01	1.15%	0.69%	1.06%	0.37%	1.21%	0.06%	1.12%	1,118.45

The Diversified Core

TABLE 6.3 (CONTINUED)
Developing Diversified Core Value at Risk

	SPY		MDY		QQQ		Gain/Loss	Gain/Loss
Date	% G/L	Wgt G/L	% G/L	Wgt G/L	% G/L	Wgt G/L	(%)	($)
02/13/01	-0.82%	-0.49%	-0.57%	-0.20%	-3.12%	-0.16%	-0.85%	(846.26)
02/14/01	-0.15%	-0.09%	0.31%	0.11%	3.89%	0.19%	0.21%	213.32
02/15/01	0.97%	0.58%	0.98%	0.34%	1.65%	0.08%	1.01%	1,006.76
02/16/01	-2.20%	-1.32%	-1.85%	-0.65%	-5.60%	-0.28%	-2.25%	(2,248.84)
02/20/01	-1.54%	-0.92%	-1.68%	-0.59%	-3.88%	-0.19%	-1.71%	(1,707.17)
02/21/01	-2.16%	-1.29%	-1.34%	-0.47%	-2.81%	-0.14%	-1.90%	(1,902.50)
02/22/01	0.15%	0.09%	-0.87%	-0.30%	-0.97%	-0.05%	-0.26%	(260.99)
02/23/01	-0.68%	-0.41%	0.11%	0.04%	0.35%	0.02%	-0.35%	(349.50)
02/26/01	2.13%	1.28%	2.62%	0.92%	1.90%	0.09%	2.29%	2,288.50
02/27/01	-0.92%	-0.55%	-2.02%	-0.71%	-6.04%	-0.30%	-1.56%	(1,563.86)
02/28/01	-1.97%	-1.18%	-1.52%	-0.53%	-3.16%	-0.16%	-1.87%	(1,871.49)
03/01/01	0.52%	0.31%	0.39%	0.13%	2.85%	0.14%	0.59%	591.88
03/02/01	-0.79%	-0.48%	0.82%	0.29%	-4.30%	-0.22%	-0.40%	(403.74)
03/05/01	0.91%	0.55%	0.00%	0.00%	1.82%	0.09%	0.64%	639.51
03/06/01	1.07%	0.64%	1.47%	0.51%	3.89%	0.19%	1.35%	1,353.50
03/07/01	0.71%	0.43%	0.64%	0.23%	0.04%	0.00%	0.66%	655.65
03/08/01	0.11%	0.07%	-0.94%	-0.33%	-1.86%	-0.09%	-0.36%	(355.29)
03/09/01	-2.96%	-1.77%	-1.76%	-0.62%	-7.01%	-0.35%	-2.74%	(2,742.95)
03/12/01	-4.28%	-2.57%	-4.10%	-1.43%	-6.21%	-0.31%	-4.31%	(4,312.56)
03/13/01	1.64%	0.99%	1.29%	0.45%	5.08%	0.25%	1.69%	1,691.70
03/14/01	-1.97%	-1.18%	-1.91%	-0.67%	-1.57%	-0.08%	-1.93%	(1,930.62)
03/15/01	0.03%	0.02%	-0.39%	-0.14%	-3.66%	-0.18%	-0.30%	(304.37)
03/16/01	-2.27%	-1.36%	-2.87%	-1.01%	-2.49%	-0.12%	-2.49%	(2,491.76)
03/19/01	2.03%	1.22%	2.97%	1.04%	5.26%	0.26%	2.52%	2,523.35
03/20/01	-2.68%	-1.61%	-2.76%	-0.97%	-6.73%	-0.34%	-2.91%	(2,912.28)
03/21/01	-1.70%	-1.02%	-1.97%	-0.69%	0.00%	0.00%	-1.71%	(1,708.80)
03/22/01	-1.02%	-0.61%	-1.03%	-0.36%	6.07%	0.30%	-0.67%	(665.88)
03/23/01	3.02%	1.81%	2.69%	0.94%	0.00%	0.00%	2.76%	2,756.15
03/26/01	1.28%	0.77%	1.25%	0.44%	-2.31%	-0.12%	1.09%	1,087.31
03/27/01	2.04%	1.23%	0.91%	0.32%	3.44%	0.17%	1.72%	1,715.76
03/28/01	-2.76%	-1.66%	-2.18%	-0.76%	-8.21%	-0.41%	-2.83%	(2,831.85)
03/29/01	0.38%	0.23%	0.00%	0.00%	-1.89%	-0.09%	0.14%	135.03
03/30/01	1.05%	0.63%	0.60%	0.21%	0.51%	0.03%	0.86%	862.94
04/02/01	-2.13%	-1.28%	-2.55%	-0.89%	-4.47%	-0.22%	-2.40%	(2,395.40)
04/03/01	-3.34%	-2.00%	-2.98%	-1.04%	-7.33%	-0.37%	-3.41%	(3,410.61)
04/04/01	0.42%	0.25%	-0.34%	-0.12%	-1.76%	-0.09%	0.04%	43.60
04/05/01	3.79%	2.27%	4.75%	1.66%	9.57%	0.48%	4.42%	4,415.57
04/06/01	-1.52%	-0.91%	-1.57%	-0.55%	-2.71%	-0.14%	-1.60%	(1,598.35)
04/09/01	1.11%	0.67%	1.63%	0.57%	2.07%	0.10%	1.34%	1,342.51
04/10/01	1.82%	1.09%	2.47%	0.87%	7.42%	0.37%	2.33%	2,330.87

(continued)

TABLE 6.3 (CONTINUED)
Developing Diversified Core Value at Risk

	SPY		MDY		QQQ		Gain/Loss	Gain/Loss
Date	% G/L	Wgt G/L	% G/L	Wgt G/L	% G/L	Wgt G/L	(%)	($)
04/11/01	0.07%	0.04%	0.35%	0.12%	3.02%	0.15%	0.31%	314.85
04/12/01	1.82%	1.09%	1.81%	0.63%	4.39%	0.22%	1.94%	1,942.23
04/16/01	-1.05%	-0.63%	-2.11%	-0.74%	-5.96%	-0.30%	-1.67%	(1,667.05)
04/17/01	1.41%	0.85%	2.00%	0.70%	2.48%	0.12%	1.67%	1,671.90
04/18/01	3.97%	2.38%	3.96%	1.39%	10.42%	0.52%	4.29%	4,291.94
04/19/01	1.33%	0.80%	0.81%	0.28%	6.04%	0.30%	1.38%	1,382.36
04/20/01	-0.92%	-0.55%	-1.10%	-0.38%	0.21%	0.01%	-0.92%	(922.14)
04/23/01	-1.82%	-1.09%	-1.24%	-0.43%	-6.71%	-0.34%	-1.86%	(1,859.01)
04/24/01	-0.54%	-0.32%	-0.43%	-0.15%	-1.88%	-0.09%	-0.57%	(567.22)
04/25/01	1.31%	0.78%	2.06%	0.72%	1.81%	0.09%	1.60%	1,596.25
04/26/01	0.45%	0.27%	1.29%	0.45%	-2.55%	-0.13%	0.59%	592.27
04/27/01	1.67%	1.00%	1.42%	0.50%	2.73%	0.14%	1.63%	1,631.19
04/30/01	-0.69%	-0.42%	0.21%	0.08%	2.21%	0.11%	-0.23%	(229.08)

Table 6.4 ranks the blended index returns from the best to the worst trading days. The VAR at the 95 percent confidence level would be the amount of the loss on the fifth worst trading day. The fifth worst return for the diversified core Account over the last 100 trading days was $2,912.

TABLE 6.4
100 Daily Hypothetical Profits and Losses from Largest Profit to Loss

Number	SPY		MDY		QQQ		Hypothetical Mark-to Market Value of Diversified Core	Change in Mark-to-Market Value of Diversified Core
	Gain/Loss (%)	Weighted Gain/Loss (%)	Gain/Loss (%)	Weighted Gain/Loss (%)	Gain/Loss (%)	Weighted Gain/Loss (%)		
1	4.80%	2.88%	3.98%	1.39%	16.84%	0.84%	$105,116.26	$5,116
2	3.79%	2.27%	4.75%	1.66%	9.57%	0.48%	104,415.57	4,416
3	3.97%	2.38%	3.96%	1.39%	10.42%	0.52%	104,291.94	4,292
4	3.28%	1.97%	3.27%	1.14%	10.99%	0.55%	103,661.31	3,661
5	3.00%	1.80%	3.98%	1.39%	7.92%	0.40%	103,586.82	3,587
6	3.48%	2.09%	2.88%	1.01%	9.38%	0.47%	103,563.28	3,563
7	3.02%	1.81%	2.69%	0.94%	0.00%	0.00%	102,756.15	2,756
8	2.03%	1.22%	2.97%	1.04%	5.26%	0.26%	102,523.35	2,523

The Diversified Core

TABLE 6.4 (CONTINUED)
100 Daily Hypothetical Profits and Losses from Largest Profit to Loss

Number	SPY Gain/Loss (%)	SPY Weighted Gain/Loss (%)	MDY Gain/Loss (%)	MDY Weighted Gain/Loss (%)	QQQ Gain/Loss (%)	QQQ Weighted Gain/Loss (%)	Hypothetical Mark-to Market Value of Diversified Core	Change in Mark-to-Market Value of Diversified Core
9	1.76%	1.05%	2.99%	1.05%	5.13%	0.26%	102,357.34	2,357
10	1.82%	1.09%	2.47%	0.87%	7.42%	0.37%	102,330.87	2,331
*								
*								
*								
91	-2.13%	-1.28%	-2.55%	-0.89%	-4.47%	-0.22%	97,604.60	-2,395
92	-2.27%	-1.36%	-2.87%	-1.01%	-2.49%	-0.12%	97,508.24	-2,492
93	-2.96%	-1.77%	-1.76%	-0.62%	-7.01%	-0.35%	97,257.05	-2,743
94	-2.76%	-1.66%	-2.18%	-0.76%	-8.21%	-0.41%	97,168.15	-2,832
95	-1.81%	-1.09%	-3.84%	-1.34%	-8.46%	-0.42%	97,146.49	-2,854
96	-2.68%	-1.61%	-2.76%	-0.97%	-6.73%	-0.34%	97,087.72	-2,912
97	-2.90%	-1.74%	-3.09%	-1.08%	-5.73%	-0.29%	96,893.78	-3,106
98	-3.26%	-1.96%	-2.74%	-0.96%	-7.65%	-0.38%	96,699.73	-3,300
99	-3.34%	-2.00%	-2.98%	-1.04%	-7.33%	-0.37%	96,589.39	-3,411
100	-4.28%	-2.57%	-4.10%	-1.43%	-6.21%	-0.31%	95,687.44	-4,313

Returns for the diversified core can be compared to those of the S&P 500 (in Column 2) to establish an indication of relative performance. In this instance, on the diversified core's fifth worst trading day, the S&P 500 was off 2.68 percent.

The VAR at the 95 percent confidence level for the diversified core blend of 65 percent SPY, 30 percent MDY, and 5 percent QQQ equals $2,912.28 while the S&P

TABLE 6.5
S&P 500 Performance Relative to Diversified Core

Number	S&P 500	Diversified Core
100	-4.28%	-4.31%
99	-3.34%	-3.41%
98	-3.26%	-3.30%
97	-2.90%	-3.11%
96	-2.68%	-2.91%
95	-1.81%	-2.85%
94	-2.76%	-2.83%

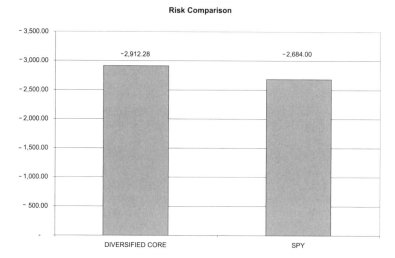

FIGURE 6.4 S&P 500 Performance Relative to Diversified Core

500 risk is $2,684.28 for a difference of only $228. While not absolutely predictive, this methodology allows a frame of reference for determining potential portfolio risk of loss. Figure 6.4 compares this dollar risk between the diversified core blend and the S&P 500 index.

At the 95 percent confidence level, the diversified core portfolio is only 8.5 percent riskier than the S&P 500. Figure 6.5 traces the daily relative performance of the diversified core and the benchmark.

The investor would now determine if this blend offers not only superior diversification, but also acceptable risk-adjusted returns relative to the benchmark S&P 500. If so, the 65/30/5 blend would become the starting point for the portfolio core.

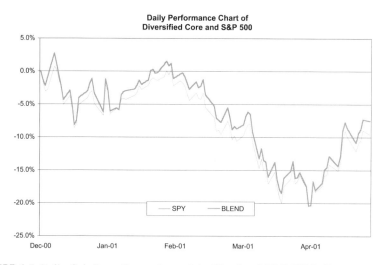

FIGURE 6.5 Daily Gain/Loss Comparison of the Blend and S&P 500 Indices

If not, the blend can be adjusted to increase or decrease VAR by setting different portfolio weights and recalculating the historical returns. The process is repeated until an acceptable figure is reached.

7 The Active Portion

As noted previously, active investment can be employed using individual stocks or other securities, including basket securities, that correspond to various sectors of the economy. In our example, the active portion of the Basket Case Portfolio can range from 25 to 50 percent of equity assets as indicated by the active/passive indicator introduced in Chapter 6. If the indicator specifies that active management is likely to outperform, the diversified core is reduced by up to 25 percent resulting in a maximum 50 percent allocation to the active component. If passive investment is currently stronger, as little as 25 percent can be apportioned to the active piece with the balance moving back into the diversified core.

The BCP active element investment menu consists of the nine S&P Select Sector SPDRs from which two or three are chosen for investment. The BCP Sector SPDR selection process is based on a combination of fundamental and technical analysis. As shown in Figure 7.1, the procedure begins with a broad examination of the business cycle, a consideration of what might happen in the financial markets, a determination of which sectors will benefit from this assessment, and finally, a technical verification of the market strength of the selected SPDRs.

THE S&P 500 SECTOR SPDRS

The sector SPDRs are well suited as active/passive investment vehicles since they provide an efficient and inexpensive way to buy or short entire market sectors. The S&P 500 is divided into four segments — industrials, utilities, transportation stocks, and financials. These four segments are further subdivided into 11 sectors consisting of 88 industries. These divisions are shown in Table 7.1.

While the S&P divides its 500 index into 11 sectors, it would not be feasible to offer 11 sector SPDRs because some would be too concentrated in big-cap issues. An IRS restriction mandates that a regulated investment company cannot invest more than 25 percent of its assets in any one stock. Thus, the 11 industries were reorganized as nine, shown in Table 7.2. Not all companies fit perfectly. For instance, General Electric is included in the industrial sector SPDR even though it has a large presence in financial services. Some of the companies in the technology SPDR, such as Boeing, Lockheed, Eastman Kodak, and BF Goodrich, are not the high-flying tech plays one might expect.

Because the index represents over 80 percent of the market value of all stocks trading on the New York Stock Exchange, it is one of the 11 leading economic indicators published by the U.S. Commerce Department.

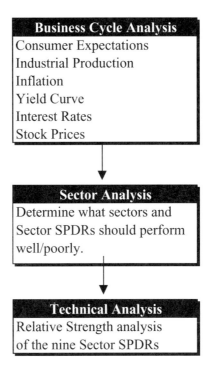

FIGURE 7.1 The Active Account Analysis Process

**TABLE 7.1
The Divisions of the S&P 500 Index**

S&P 500 Index

4 Segments
11 Sectors
88 Industries
500 Companies

STEP ONE — EXAMINING THE BUSINESS CYCLE

The economy is cyclical in nature, regularly moving through periods of expansion and contraction. This recurring pattern of recession and recovery is referred to as the **business cycle.** In simplest form, illustrated in Figure 7.2, the business cycle begins in recession, moves slowly to accelerating growth, and eventually reaches a peak, falters, deteriorates, and falls back to a recession to complete the cycle.

This repetitiveness affords a limited level of predictability that can be exploited. The most basic task before the investor is to determine whether the economy is improving or deteriorating. Once established, a more responsive investment strategy can be implemented.

TABLE 7.2
The Active Portion Investment Menu

Select Sector SPDR	Ticker
Consumer Services	XLV
Basic Industries	XLB
Utilities	XLU
Cyclical/Transportation	XLY
Consumer Staples	XLP
Technology	XLK
Industrial	XLI
Financial	XLF
Energy	XLE

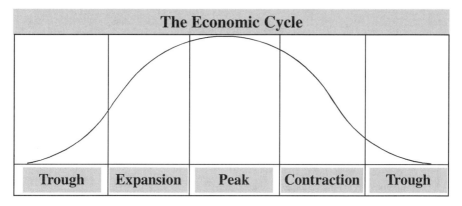

FIGURE 7.2 The Economic Cycle

By examining past cycles and present data, the investor creates a frame of reference for viewing the unfolding of present and future events. To assist in this process, the National Bureau of Economic Research (NBER) has created a set of three cyclical indicators to help forecast, measure, and interpret short-term fluctuations in economic activity. There are 11 leading, four coincident, and seven lagging indicators. As their names suggest, they either anticipate, move with, or move slightly behind the rest of the economy.

While the sum of all these measures helps to identify the peaks and troughs of the business cycle, some are more useful than others. Far fewer are sufficient to reach an acceptable conclusion. An observance of consumer expectations, industrial output, inflation, interest rates, the slope of the yield curve, and stock prices is generally sufficient in determining the present position in the business cycle.

Stock prices, or the level of the S&P 500, are of particular importance. Equities generally follow a pattern that is similar to but anticipating that of the business cycle. Specifically, as shown in Figure 7.3, stocks rise about five months prior to the onset of expansion, and fall about six months before the onset of an economic contraction.

TABLE 7.3
NBER Economic Indicators

Leading Indicators

1. Average weekly hours of production
2. Average weekly initial claims for unemployment insurance
3. Manufacturers' new orders
4. Vendor performance
5. Contracts and orders for land and equipment
6. New private housing units authorized by local building permits
7. Change in manufacturers' unfilled orders
8. Change in sensitive materials prices
9. Stock prices, 500 common stocks
10. Money supply (M2)
11. Index of consumer expectations

Coincident Indicators

1. Employees on nonagricultural payrolls
2. Personal income less transfer payments
3. Industrial production
4. Manufacturing and trade sales

Lagging Indicators

1. Average duration of unemployment
2. Ratio of trade inventories to sales
3. Change in index of labor cost per unit of output
4. Average prime rate charged by banks
5. Commercial and industrial loans outstanding
6. Ratio of consumer installment credit outstanding to personal income
7. Change in consumer price index for services

Many industry groups demonstrate established patterns of outperforming the overall stock market during specific phases of the economic cycle and underperforming during others.

Companies are usually identified as being either early-, mid-, or late-cycle performers, or being defensive or interest-sensitive. In order to identify when certain sectors may perform best or worst, the economic cycle can be broken down into the five phases illustrated in Figure 7.4: three expansion (early, middle, and late) and two contraction (early and late).

Table 7.4 offers a tabular presentation of the six suggested indicators, their duration, characteristics, and the conventional stock market reaction during a typical business cycle.

Table 7.5 provides a useful abbreviation of the previous table that can be used as a checklist in gauging the variables. Review each characteristic to discern any emerging or existing patterns.

The Active Portion 87

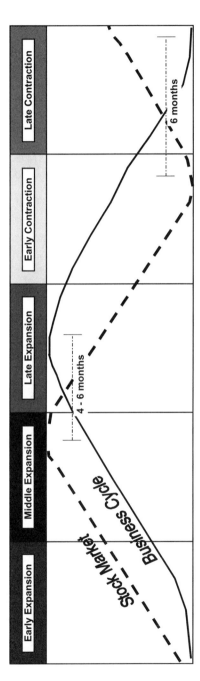

FIGURE 7.3 The Market as a Leading Indicator

	Early Expansion	Middle Expansion	Late Expansion	Early Contraction	Late Contraction
Consumer Expectations	Rising Sharply	Leveling Off	Declining	Falling Sharply	Reviving
Industrial Production	Flat to Rising Modestly	Rising Sharply	Flattening Out	Declining	Decline Diminishing
Interest Rates	Bottoming Out (leads inflation)	Rising Modestly	Rising Rapidly	Peaking	Falling
Yield Curve	Steep	Moderate	Flattening Out	Flat (and sometimes inverted)	Rising Again
Sector	Transportation Technology	Services Capital Goods	Basics Materials Energy	Consumer Staples	Utilities Financials Consumer Cyclical
Sector SPDR	Cyc/Transportation XYL Technology XLK	Consumer Services XLV Industrial XLI	Basic Industrial XLB Energy XLE	Consumer Staples XLP	Utilities XLU Financial XLF Cyc/Transportation XYL

FIGURE 7.4 The Five Phases of the Business Cycle

TABLE 7.4
A Tabular Representation of the Business Cycle

Early Expansion

Duration:	First third of economic expansion, or about 17 months on average.
Consumer expectations:	Rising sharply.
Industrial production:	Flat to rising modestly.
Inflation:	Continuing to fall.
Interest rates:	Bottoming out (leads inflation).
Yield curve:	Steep.

Middle Expansion

Duration:	Second third of economic expansion, or about 17 months on average.
Consumer expectations:	Leveling off.
Industrial production:	Rising sharply.
Inflation:	Bottoming out.
Interest rates:	Rising modestly.
Yield curve:	Moderate.

Late Expansion

Duration:	Last third of economic expansion, or about 17 months on average.
Consumer expectations:	Declining.
Industrial production:	Flattening out.
Inflation:	Rising modestly and beginning to be of concern to investors and the Fed.
Interest rates:	Rising rapidly due to Fed policy (as well as the supply and demand of capital) to combat inflation.
Yield curve:	Flattening out (short rates rising as the Fed combats inflation, whereas long rates may be falling as they reflect future inflationary expectations).

Early Contraction

Duration:	First half of economic contraction, or about 6 months on average.
Consumer expectations:	Falling sharply.
Industrial production:	Declining.
Inflation:	Rising less strongly.
Interest rates:	Peaking.
Yield curve:	Flat (and sometimes inverted — short rates are higher than long rates).

Late Contraction

Duration:	Final half of economic contraction, or about 6 months on average.
Consumer expectations:	Reviving.
Industrial production:	Decline diminishing.
Inflation:	Flat to declining.
Interest rates:	Falling.
Yield curve:	Rising again.

TABLE 7.5
Verification Checklist

	Early Expansion	Middle Expansion	Late Expansion	Early Contraction	Late Contraction
Duration (Months)	☐ Average 17	☐ Average 17	☐ Average 17	☐ Average 6	☐ Average 6
Consumer Expectations	☐ Rising sharply	☐ Leveling off	☐ Declining	☐ Falling sharply	☐ Reviving
Industrial Production	☐ Flat to rising modestly	☐ Rising sharply	☐ Flattening out	☐ Declining	☐ Decline diminishing
Inflation	☐ Continuing to fall	☐ Bottoming out	☐ Rising modestly	☐ Rising less strongly	☐ Flat to declining
Yield Curve	☐ Steep	☐ Moderate	☐ Flattening out	☐ Flat to inverted	☐ Rising again
Interest Rates	☐ Bottoming out	☐ Rising modestly	☐ Rising rapidly	☐ Peaking	☐ Falling
Stock Prices	☐ Rising modestly	☐ Rising faster	☐ Peaking	☐ Falling sharply	☐ Rising again

Consumer Expectations

Consumer expectations is one of the leading economic indicators (LEI) and leads changes in consumer spending. Optimism leads to increased consumption and thus production, while pessimism leads to reduced spending and output. While these and other aspects of demand are tied to the business cycle, changing demographics, fashions, prices, or cultural concerns also influence demand. It is released in the fourth week of each month. Strong numbers may move bonds lower on inflationary fears but stocks could move higher. Weaker than expected numbers usually mean higher bond prices and lower stocks.

Industrial Production

Industrial production measures the physical quantity of goods produced by the manufacturing, mining, and utility sectors. It is considered an excellent coincident economic indicator, accurately measuring the current state of the economy. It is released around the 15th of each month with data corresponding to the prior month.

Inflation

Inflation is measured by two primary standards, the Producer Price Index (PPI) and the Consumer Price Index (CPI). The PPI is reported around the second week of the month (usually Thursday or Friday) and the CPI comes about one week later. The PPI will usually turn higher before an upswing in consumer prices. Food and energy prices are removed in order to gauge "core" inflation. Inflationary numbers will hurt bond prices. Stocks could initially rally over the short run. If the Fed tightens, equity values will eventually start to decline. Lower inflation could rally both bonds and, because of the lower yields, stocks.

Interest Rates

The Federal Reserve board employs four primary tools in their quest to influence the money supply, credit conditions, and interest rates. They can:

1. Set bank reserve requirements
2. Move the discount rate
3. Change margin requirements
4. Buy or sell government securities in the open market.
5. Move the Fed funds target rate.

Reserves are the required percentage of deposits that banks and thrifts must hold in cash or in deposits at the Federal Reserve. An increase in reserve requirements means few funds for lending. Federal Funds are reserve balances above those required that are maintained by commercial banks in the Federal Reserve System. These excess reserves are available for lending to other banks in need of reserves. Such loans are normally made on a single day basis. The Fed Funds Rate is the rate of interest on overnight loans of excess reserves among commercial banks.

The discount rate is the rate that the Federal Reserve will charge member banks that borrow reserves. The Fed is the lender of last resort. The expression comes from the early days of the Fed when the transaction involved discounting bank loan portfolios that were brought to the Fed's "Discount Window" as collateral.

Margin requirements set the minimum portion of a new security purchase that an investor must pay for in cash. The lowest point was in 1930 when the reserve was set at 10 percent. The highest rate was in the early 1950s at 60 percent.

The purchase and sale of government securities in the open market by the Fed is useful in supplying or contracting liquidity or money supply.

YIELD CURVE

Interest rates are plotted on a yield curve, which is a graphical depiction of yields offered by three-month T-bills to 30-year Treasury bonds. A normal curve slopes upward because investors demand higher yields on long-term bonds to make up for their greater risk. The spread between short-term bills and long-term bonds usually runs about 2 percent.

In the early stages of economic expansion, the yield curve rises sharply and the spread swells to 3 percent or more. Investors are beginning to foresee inflationary pressure from accelerating growth. Preceding an economic contraction, on the other hand, the yield curve will tighten to 1 percent or less. Occasionally the yield curve becomes inverted, with short-term bonds yielding more than long-term bonds. This occurs when investors anticipate the slower growth and reduced inflationary pressure of an approaching recession.

STOCK PRICES

The stock market, as measured by the S&P 500 index, is a leading economic indicator. This is sensible since stock prices are forward-looking predictors of future profitability. Stock prices generally follow a pattern similar to, but anticipating, the business cycle by rising about five months prior to the onset of expansion and falling about six months before the onset of economic contraction. Many industry groups generally demonstrate established patterns of outperforming the overall stock market during specific phases of the economic cycle and underperforming during others.

Another useful stock market/business cycle indicator is the performance of small and mid-cap stocks. These companies typically begin a prolonged period of outperformance as the overvalued S&P 500 reaches a peak prior to a business cycle peak. Smaller companies are routinely neglected during expansionary periods, and because they represent greater growth potential at these times, they attract investor attention.

STEP TWO — SECTOR ANALYSIS

SECTOR ROTATION

The trend of companies to roll in and out of favor is described as *sector rotation*. Successful investing in the active account is determined by how accurately sector

The Active Portion

rotation can be anticipated. The objective is to identify the current phase in the economic cycle and thus anticipate the timing of the succeeding phase.

While the active account rotates portfolio holdings by buying those Sector SPDRs projected to outperform and selling those that have already peaked, some care must be applied. The stock market is not always completely accurate at forecasting economic downturns. For example, in 1980, the stock market peaked a full month *after* the start of the economic downturn. In November 1968, it peaked 13 months ahead of the onset of the contraction. However, on average the market typically anticipates an economic contraction by an average of six months.

The stock market is more successful at anticipating the beginnings of economic expansion. Over the past five expansionary periods the market bottomed between three and six months with an average record of 4.5 months prior to an economic revival.

Anticipating sector or industry performance within particular phases of the economic cycle is the basis for the active account management process. Being able to accurately perceive market turns in anticipation of changing economic conditions helps the power investor gain a competitive advantage. The goal is to accumulate industries projected to perform well while avoiding those expected to suffer.

Let's revisit the traditional business cycle using a different graphical representation, shown in Figure 7.5 to notice how the sectors and sector SPDRs would normally

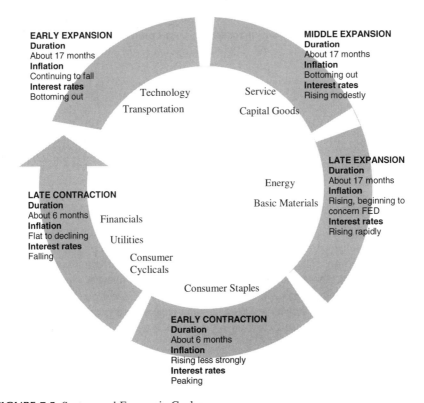

FIGURE 7.5 Sectors and Economic Cycles

TABLE 7.6
Technology Sector SPDR (XLK)

Technology
Communication equipment/manufacturers
Computer software and services
Computer systems
Electronics (defense)
Electronics (instrumentation)
Electronics (semiconductors)
Office equipment and supplies
Photography/imaging
Telecommunications (long distance)

be expected to perform. The nine sector SPDRs are on the interior of the circle associated with the phase in which they normally outperform the broader market.

EARLY EXPANSION

As demand begins to rise, companies seek to improve productivity through technology, so computer and other high-tech stocks also prosper. As the economy expands more, transportation stocks will thrive as more and more goods are delivered.

Technology

Technology stocks are most popular during early to mid stages of an economic expansion. Technology stocks can be cyclical to the degree that they depend on capital spending and business or consumer demand. However, they may also have long-term growth potential as technological products find broader applications and as new technologies are developed. Industries in the technology sector SPDRs are listed in Table 7.6.

Transportation

Railroads and other surface carriers tend to react early to a pickup in the economy. Airlines are subject to cyclical fuel costs, usage versus capacity, and competitive pressures on airfares. The coupling of the cyclical and transportation companies into one sector SPDR, shown in Table 7.7, unfortunately complicates a pure play here. However, combining the two slightly out-of-sync sectors is not particularly harmful. The consumer cyclicals are slightly ahead of the transportation companies in the sector rotation schedule, but are sufficiently close to avoid a tug of war investment effect.

Consumer Cyclicals

Stocks in this category, also shown in Table 7.7, include durable and nondurable products and are sensitive to interest rates. As interest rates move lower, consumers

---

**TABLE 7.7
Cyclical/Transportation Sector and Consumer Cyclical Sector SPDRs**

Cyclical/Transportation Sector SPDR (XLY)
Transportation
Airlines
Railroads
Truckers
Transportation (Misc.)

Consumer Cyclical Sector
Auto parts after market
Automobiles
Broadcast media
Building material
Entertainment
Hardware and tools
Homebuilding
Hotel-motel
Household furnishings and appliances
Leisure time
Manufactured housing
Publishing
Publishing (newspapers)
Restaurants
Retail (department stores)
Retail (general merchandise)
Retail (specialty — apparel)
Shoes
Textiles
Toys

---

are motivated to purchase more. Investors typically seek them out in the late stages of contraction, but their relative strength can remain strong throughout the beginning phase of expansion.

MIDDLE EXPANSION

As the economy reaches the middle phase of expansion, service stocks, like providers of temporary workers, rise as companies struggle to meet increased demand but remain reluctant to add to permanent overhead costs.

Capital Goods

Capital spending tends to increase midway through the business cycle, as the economy is heating up and higher demand for products leads companies to expand their

TABLE 7.8
Industrial Sector SPDR (XLI)

Capital goods sector
Aerospace/defense
Conglomerates
Electrical equipment
Engineering and construction
Heavy duty trucks and parts
Machine tools
Machinery (diversified)
Manufacturing (diversified industries)
Pollution control

production capacity. Demand in global export markets is key for agricultural equipment, industrial machinery, and machine tools. The industrial sector SPDR is described in Table 7.8.

Consumer Services

Not all components of the consumer services SPDR, shown in Table 7.9 fit neatly into the middle expansion period. However, publishing, lodging and gaming, and prepared foods are very good matches, while the remaining industries still perform well during this period in the cycle.

LATE EXPANSION

Finally, when the expansion is in full throttle, companies hire a few more workers but spend more money on capital goods. That makes stocks in machine tools, energy, and basic materials like steel the preferred investment.

Energy

Stocks in the energy sector, listed in Table 7.10 tend to be popular with investors late in the business expansion. This category includes large integrated international companies, domestic exploration companies, and energy services companies. Each

TABLE 7.9
Consumer Services SPDR (XLV)

Entertainment
Publishing
Prepared foods
Medical services
Lodging and gaming

TABLE 7.10
Energy Sector SPDR (XLE)

Energy
Oil and gas drilling
Oil (domestic integrated)
Oil (exploraton and production)
Oil (international integrated)
Oil well equipment and services

industry has its own dynamics, but ultimately the worldwide supply and demand picture for energy drives all prices. Political events historically have had a major impact on these industries.

Basic Industry

The global economic supply and demand picture affects basic material company price behavior. Profits are driven by high capacity utilization and strong market demand for products. Therefore, these stocks (see Table 7.11) tend to be popular with investors late in an economic expansion.

Investors also often turn to precious metals and the stocks of companies that mine and process them late in the expansion cycle. The primary motivation is a sense of inflationary protection. The rapid movement into and out of these stocks generates high volatility. However, precious metal companies are not a large component of the basic industries sector SPDR.

EARLY CONTRACTION

As the expansion matures, prolonged rapid growth causes interest rates to rise; the market averages begin to decline, and investors turn to more defensive issues. These

TABLE 7.11
Basic Industries Sector SPDR (XLB)

Basic materials sector
Aluminum
Chemicals
Chemicals (diversified)
Chemicals (specialty)
Containers (metal and glass)
Gold mining
Metals (miscellaneous)
Paper and forest products
Steel

TABLE 7.12
Consumer Staples Sector SPDR (XLP)

Consumer staples
Beverages (alcoholic)
Beverages (soft drinks)
Cosmetics
Distributors (consumer products)
Foods
Healthcare (diversified)
Healthcare (drugs)
Healthcare (miscellaneous)
Hospital management
Household products
Housewares
Medical products and supplies
Retail (drug stores)
Retail (food chains)
Tobacco

are companies that are more independent of the business cycle. For instance, demand for tobacco products is not tied to the state of the economy. Health care, medical services, and food companies are necessities and show little sensitivity to business conditions. Entertainment stocks are also fairly attractive at these times as consumers tend to substitute movies for more expensive sources of entertainment.

Consumer Staples

Consumer staples are most attractive when moving from late expansion to early contraction. Consumer staples, listed in Table 7.12, include the noncyclical companies such as tobacco, beverages, food, and cosmetics. These companies usually experience steady demand and are not sensitive to business cycle changes.

Healthcare stocks are also considered defensive since companies in this sector are little affected by economic variability. Pharmaceutical firms, HMOs, biotechnology firms, and medical equipment providers make up the healthcare industries. Pharmaceutical companies are highly competitive and are affected by the pace of FDA approvals, patent lives, and the strength of the R&D pipelines. Biotechnology firm valuations are tied to R&D pipelines and investor perceptions.

Consumer Cyclicals

Durable and nondurable goods become more attractive as the lower interest rates begin to spur consumer activity. The consumer cyclical/transportation SPDR (XLY) becomes the vehicle of choice here.

TABLE 7.13
Financial Sector SPDR (XLF)

Financials
Life insurance
Major regional banks
Money center banks
Multiline insurance
Personal loans
Property-casualty insurance
Savings & loan companies
Financial (miscellaneous)

Financial Services
Insurance brokers
Specialized services
Specialty printing

LATE CONTRACTION

Finally, during the late contraction phase, the financial issues become attractive as the Fed's aggressive easing of interest rates pulls the economy out of recession and creates lower interest rates that generate increased consumption and production. At this point the consumer cyclicals take off and the sector rotation progression begins anew.

Financials

Investors target financials most often in the mid to late stages of an economic contraction. Banks and savings and loans carry deposits as liabilities. Fed-inspired low interest rates help these institutions free more capital for lending, or asset generation. Housing-related companies typically respond well to falling interest rates while the non-mortgage-dependent banks prosper by commercial and consumer loan growth. These latter firms come into favor during the middle of the late contraction cycle. Table 7.13 shows the financial sector SPDR.

Utilities

Electric utility companies tend to perform well in an environment of declining interest rates because the large debt financing costs incurred to build their infrastructures can now be serviced with much less pressure.

As the promise of economic growth increases, telecommunication stocks become more attractive due to rising demand for telephone services and ancillary deregulated products.

Natural gas stocks are also positively affected by future growth trends anticipated during the late contraction phase. Growth in demand is correlated with

TABLE 7.14
Utilities Sector SPDR (XLU)

Electric companies
Natural gas
Telephone

industrial production and housing starts. Table 7.14 lists the content of the utilities sector SPDR.

A summary of the various sectors and sector SPDRs likely to outperform the general market during the five phases of the classic business cycle is presented in Table 7.15.

STEP THREE — TECHNICAL ANALYSIS

An analysis of the business cycle is useful in determining the current position in the economy and is suggested as a minimal analytical investment selection process. However, two critical questions remain before the final selection of the BCP's active equity components. First, has the analysis of the business cycle been correct? Second, will the historic business cycle pattern emerge, or will it be different this time?

If the factors propelling the economy are the traditional cyclical ones, the market has probably already priced them into the equation. This necessitates finding good entry points on pullbacks for any purchases. But what if this time, things are different? For example, has technologically driven rising productivity eliminated the expected inflationary pressures of increased demand? If so, has the market anticipated or adjusted to these new factors or elements? If it has, will a traditional rotation

TABLE 7.15
Business Cycle and Relative Sector Performance

Economic Phase	Sectors Expected to Outperform	Sector SPDR
Early expansion	Transportation	Cyclical/transportation (XLY)
	Technology	Technology (XLK)
Middle expansion	Services	Consumer services (XLV)
	Capital goods	Industrial (XLI)
Late expansion	Basic materials	Basic materials (XLB)
	Energy	Energy (XLE)
Early contraction	Consumer staples	Consumer staples (XLP)
Late contraction	Utilities	Utilities (XLU)
	Financials	Financials (XLF)
	Consumer cyclicals	Cyclical/transportation (XLY)

The Active Portion

to energy or basic materials be warranted, or should the portfolio remain with capital goods and services sectors?

These difficult questions are both objective and subjective. A definitive position is difficult to achieve. As an example, refer to Table 7.16 for an assessment of the economic conditions prevalent as of mid-2001.

The pattern seems to indicate that the economy is in the middle to late contraction period. Consumer expectations are just beginning to drop as industrial production declines and layoffs build. Inflation is flat to declining thanks to high productivity and lack of global pricing power. The yield curve is moderate, but threatening to steepen as the long end of the yield curve begins to sense a return of inflationary forces. Interest rates are still falling, but beginning to find a bottom. Stock prices are mixed as the market remains in a two-year-old trading range after a volatile correction in technology and momentum-type companies. Obviously, no clear-cut picture emerges from this review.

Something is different this time. Despite falling corporate margins and rising layoffs, the consumers continue to open their wallets, offsetting the decline in corporate spending and saving the economy from slumping into a recession. If this national shopping spree should end, the unpleasant consequence would be negative for both the economy and the stock market.

Table 7.16 seems to indicate that the financial (XLF) and cyclical/transportation (XLY) sector SPDRs that characteristically outperform in the late contraction period would be the preferred investment choices.

Before the final investment decision is made, however, it is recommended that the stock market's forward-looking predictive power be considered. The efficient market hypothesis states that at any given point in time, stock prices reflect all currently available information. One of the best ways to measure the market tide, and thus capitalize on its self-correcting characteristic, is through relative strength analysis.

Figure 7.6 presents the 13-week moving-average percentile relative strength for cash, the S&P 500, and each of the nine sector SPDRs that existed concurrently with the economic factors observed in Table 7.16. The nine SPDRs are grouped by their traditional business cycle phases of outperformance. The relative strength numbers are based on a range of 1 to 99, with the top SPDR ranked in the 99th percentile and the bottom SPDR rated at a percentile of one.

It is apparent that the market is pricing a recovery and predicting a **future move** into the **early** stages of expansion. This is indicated by the two highest relative strength securities, the cyclical/transportation (XLY) and consumer services (XLV) sector SPDRs. The former is a traditional late contraction/early expansion sector performer while the latter is both a safe harbor, and a middle expansion performer. Although not visible on this snapshot, the financial SPDR's (XLF) relative strength has been declining over the past three months. Cash is currently very low on a relative percentage basis. According to the chart, now is the time to be anticipating a recovery with a possible gradual accumulation program into early and middle expansion sector SPDRs.

TABLE 7.16
A Sample Assessment

	Early Expansion	Middle Expansion	Late Expansion	Early Contraction	Late Contraction
Consumer Expectations	☐ Rising sharply	☑ Leveling off	☐ Declining	☐ Falling sharply	☐ Reviving
Industrial Production	☐ Flat to rising modestly	☐ Rising sharply	☐ Flattening out	☑ Declining	☐ Decline diminishing
Inflation	☐ Continuing to fall	☐ Bottoming out	☐ Rising modestly	☐ Rising less strongly	☑ Flat to declining
Yield Curve	☐ Steep	☑ Moderate	☐ Flattening out	☐ Flat to inverted	☐ Rising again
Interest Rates	☐ Bottoming out	☐ Rising modestly	☐ Rising rapidly	☐ Peaking	☑ Falling
Stock Prices	☐ Rising modestly	☐ Rising faster	☐ Peaking	☐ Falling sharply	☑ Falling

The Active Portion 103

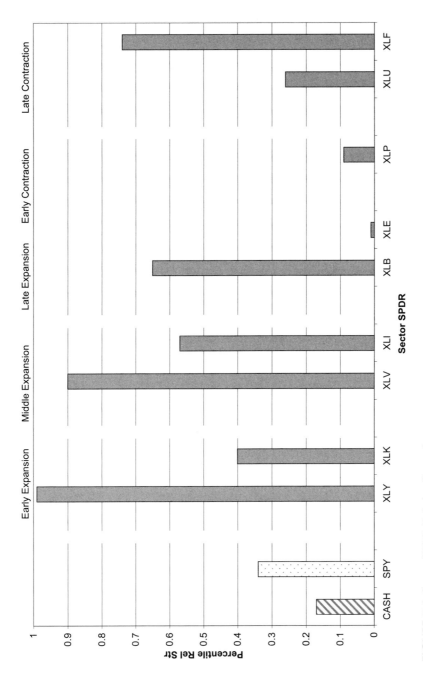

FIGURE 7.6 Sector SPDR Relative Strength

CONCLUSION

Section 2 has presented an introductory example of a self-contained equity investment process that combines active and passive portfolio management through the use of fundamental and technical tools. While by no means exhaustive, it is imminently usable and is offered as a model for the development of future, more complete applications that take advantage of the diversification, liquidity, low-cost, tax-efficiency, and marginability of the basket security.

Section III

International Investing and Looking Forward

8 Exchange-Traded Funds on International Markets

Country basket securities were first introduced in 1996, with a primary emphasis on developed stock markets. Of the 17 global markets initially selected, 14 were in the Morgan Stanley Capital International 20-country EAFE index (Europe/Australasia/Far East). This first incarnation, registered as open-end investment companies and shown in Table 8.1, was collectively referred to as "World Equity Benchmark Shares," or WEBS. Their name was later changed to iShares MSCI.

Each iShares MSCI portfolio corresponds to a basket of stocks calculated to track the similar Morgan Stanley Capital International country index. The iShares MSCI portfolios differ from SPDRs, Triple Q, and Diamonds in that they do not invest in every stock in their respective benchmarks. Instead, in seeking to replicate their country indexes, they use a representative sample method using portfolio optimization. Correlations to the underlying MSCI indexes are generally close, averaging 0.90. All iShares MSCI baskets reflect a large-cap bias because each of the underlying country indexes is value-weighted. Since the portfolios hold a relatively small number of stocks, index component changes resulting from corporate mergers and spin-offs generate moderately high turnover rates similar to those of the sector SPDRs.

Small country iShares MSCI, like sector SPDRs, can be top-heavy with more than 20 percent invested in the top company. As such, they are classified as "non-diversified" funds under the Investment Company Act of 1940. The three largest holdings in the Austrian and Belgium funds recently accounted for nearly 50% of assets. Larger funds, such as the Japan fund are less heavily concentrated, with the top three companies generally falling below eighteen percent of assets.

The success of iShares country funds is evidenced by their rapid growth in assets and trading volume during a period when many closed-end country funds have open-ended or even liquidated. Total net assets of all 17 portfolios amounted to nearly $1.5 billion as of March 31, 2001.

With $558 million in assets, iShares MSCI Japan is the largest. A few of the funds provide access to countries that do not have single-country funds available in the U.S. These include Belgium and Sweden. Expense ratios on all iShares MSCI are 0.84 percent, which, while high, is far below those of most comparable actively managed single-country funds.

Like all single-country investments, including both closed and open-end funds, iShares MSCI can be extremely volatile and should only be allocated a small portion of an investment portfolio. Single-country funds are most applicable for larger portfolios seeking a selected country exposure and can supplement open- and closed-end regional and international fund holdings.

TABLE 8.1
The International iShares

Security	Ticker
iShares MSCI Australia Index Fund	EWA
iShares MSCI Austria Index Fund	EWO
iShares MSCI Belgium Index Fund	EWK
iShares MSCI Brazil Index Fund	EWZ
iShares MSCI Canada Index Fund	EWC
iShares MSCI EMU Index Fund	EZU
iShares MSCI France Index Fund	EWQ
iShares MSCI Germany Index Fund	EWG
iShares MSCI Hong Kong Index Fund	EWH
iShares MSCI Italy Index Fund	EWI
iShares MSCI Japan Index Fund	EWJ
iShares MSCI Malaysia Index Fund	EWM
iShares MSCI Mexico Index Fund	EWW
iShares MSCI Netherlands Index Fund	EWN
iShares MSCI Singapore Index Fund	EWS
iShares MSCI South Korea Index Fund	EWY
iShares MSCI Spain Index Fund	EWP
iShares MSCI Sweden Index Fund	EWD
iShares MSCI Switzerland Index Fund	EWL
iShares MSCI Taiwan Index Fund	EWT
iShares MSCI United Kingdom Index Fund	EWU
iShares S&P Europe 350 Index Fund	IEV
iShares S&P/TSE Index Fund	IKC

REASSESSING INTERNATIONAL INVESTING

It has been nearly 30 years since the introduction of the idea that an internationally diversified equity portfolio can enhance returns while lowering risk. Since the mid-1970s the risk reduction and return enhancement factors of a diversified global portfolio have been circulated from both the academic and professional investment communities. However, how has the strategy actually worked out? Does international equity portfolio diversification truly increase return per unit of risk for U.S., or other nationally based investors?

Recent studies unfortunately provide mixed results. For instance, some studies have concluded that the addition of an international component to an equity portfolio does not materially increase expected returns, nor afford effective risk reduction. The evidence has shown that the increasing globalization of equity markets has created return variances and correlations with the Unites States that are simply too high to offer sufficient benefit. For example, from 1970 through 1994, the Europe, Australia, Far East Index (EAFE), an index that measures the market returns for developed economies, outperformed the S&P 500, but only because foreign currencies outperformed the dollar. The currency-adjusted returns of the two indexes were virtually the same. But what about emerging countries, do they offer diversification benefits for a portfolio?

The goal of international diversification is to reduce risk and enhance return. Correlations have been rising over the 1990s as markets have become more integrated and cross-border mergers have flourished. Currently, non-U.S.-developed market correlation with the S&P 500 is very high, at over 0.90. In addition, during periods of global distress, when low correlations are most needed, correlations have proven to move higher because all markets are affected negatively.

One of the contributors to a lower correlation in the past was currency movement. Foreign investments are denominated in non-U.S. currencies. As the values of those currencies change relative to the dollar, the value of foreign investment will rise and fall in dollar terms. The convergence of monetary policy around the world has helped to mitigate the swings in currency among the most developed nations and limit the contribution of currency fluctuation to overall asset class correlation.

Support of international investing based on the low correlation with U.S. markets has been weak in recent years, prompting a need for a reexamination of the benefits of investing outside U.S. borders. While domestic and foreign developed markets converge, the correlations between domestic and emerging markets have not increased significantly over time. For instance, when the U.S. market is down, the correlation with emerging foreign markets actually falls, leading to the conclusion that diversification does work.

But can this apparent favorable situation be tested? In an attempt to determine the diversification characteristics of emerging markets, historical returns have been shown to be essentially uncorrelated with returns from a future period and thus are not a good indicator of future international returns.

It is known that cross-market correlations rise through time as economic and financial integration increases. Hence, the more developed the emerging market is, the more correlation arises with the developed markets. Less mature emerging markets without these economic ties have not shown the tendency for rising correlations.

This leads, in essence, to the belief that the benefits of international diversification depend on the time period assessed and the stage of development of the particular market. During a 10-year period between 1985 and 1995, emerging markets provided superior diversification benefits by enhancing returns and lowering risk. However, other holding periods showed different results.

One fact in favor of maintaining some international exposure is that today more than half of the available stock market opportunities can be found outside the United States. Most industries today are worldwide in scope. A portfolio cannot be fully diversified in many industries, such as automobiles, consumer products, or technology, to name just a few, without including some exposure to key European and Japanese companies.

As the world becomes more integrated, a new way to approach international investing is warranted. A successful strategy is to approach foreign investing like domestic by breaking down investments according to style and market capitalization. This blend thus focuses on the mix of international assets between growth and value styles along with capitalization.

As the global capital markets continue to evolve, investors will have to constantly alter and improve their approach to investing. The required changes going forward will focus on better ways to manage risk while enhancing portfolio return.

The essential point at this stage is that international diversification can be most beneficial when distinguishing between developed and emerging markets, style, and market capitalization. In addition, the investor must be possessed of a strong ability to pick the right country at the right time. In fact, selecting the right market at the right time becomes the essence of a successful diversification strategy when utilizing foreign securities. The following section introduces the reader to one useful momentum strategy for investors in iShares country baskets.

A MOMENTUM STRATEGY FOR INTERNATIONAL EQUITY MARKETS

Accuracy in country selection timing and reallocation is critical for the global investor. A recent study by Kalok Chan, Allaudeen Harneed, and Wilson Tong entitled *Profitability of Momentum Strategies in the International Equity Markets*, (*Journal of Financial and Quantitative Analysis* 35, no. 2, June 2000), analyzed the usefulness of momentum strategies in country selection using market index rather than individual stock returns. Their results, described in the *Journal of Financial and Quantitiative Analysis,* indicate that momentum strategies do provide positive returns that are statistically and economically significant, and higher than those obtained from a buy-and-hold approach. The index approach is especially interesting for the iShares country basket investor.

The strategy assumed both long and short positions in stock market indexes, and converted returns into U.S dollars. This did cause both market performance and currency exchange rates to play a role in performance, but the currency effect was limited to 10-percent for a one-week holding period and 20 percent for a two-week horizon.

The study created zero-investment portfolios by going long in countries with above-average returns and short in countries with below average returns in period *t-1*. Country weights were proportional to the difference between individual market returns and the cross-sectional averages for the previous period. Thus, the best performing countries in period *t-1* received the highest weight in period *t*, and vice versa. Total portfolio return was estimated over 1to 26-week holding periods.

According to the study, adjusting portfolio weights in response to market performance in the previous week provided statistically significant positive returns of 0.48 percent a week for two-week holding periods and 0.25 percent a week for four-week holding periods. Results were calculated as the difference between the weighted-average returns in countries held long and short. Managing portfolios in this manner would be expected to outperform a buy and hold approach by approximately 1 percent monthly.

9 The Future

FIXED-INCOME BASKETS

During the year 2000, dozens of exchange-traded funds were launched. The appeal of low-cost, tax-efficient vehicles that trade like stocks, with prices changing from trade to trade was increasingly popular. Although similar in convenience to open- and closed-end mutual funds, the continuous trading made them highly useful to both traders and long-term investors.

But one problem remained: Up until now all ETFs invested solely in equities. There was no ETF that allowed investment in a diversified bond portfolio at the same low cost and with the same portfolio advantages. Bond investors were simply out of luck.

Recent announcements have promised to remedy this deficiency. Two firms have started the necessary process to offer ETFs based on U.S. government bond indices.

ETF innovator Barclays Global Investors Services hopes to offer fixed income ETFs in 2001, when they issue five bond ETFs based on Lehman Bros. Bond indices. The new bond ETFs will fall under the Barclays iShares designation, as its other ETFs do. They are expected to include iShares 1–3-Year Treasury, iShares 7–10-Year Treasury, and iShares 20-Year Treasury Index Funds. A broad-based iShares Treasury Fund will be based on the average of all these indices. The fifth offering will be an iShares Government/Credit Bond Index Fund based on the combined Lehman Bros. Government/Corporate bond index.

The other entrant into the bond ETF arena is Nuveen Investments. The company has filed with the SEC to sell bond ETFs too. Nuveen typically offers actively managed, tax-free bond funds. But the new Nuveen ETFs will track the Treasury Index established by Ryan Labs Inc., a New York-based research firm focusing on indices and bonds.

Nuveen plans to call these ETFs "Fixed Income Trust Receipts." The exchanged-traded funds will track one-, two-, five-, and 10-year Treasury indices and a STRIPs index that tracks zero-coupon bonds.

Bond ETFs will help to round out the ETF product line. The biggest market will most likely be the individual investor. Long-term investors will find no easier way to increase exposure to fixed income securities while traders will have the advantage of moving into and out of various bond sectors quickly and efficiently.

Financial advisors and brokers should also be attracted to bond ETFs. Many have a number of accounts with relatively small amounts to spend for a diversified portfolio of individual bonds. Such limitations do not allow the use of individual Treasuries without a lot of expense. Bond ETFs will give these investors much more flexibility at a lower cost than comparable bond mutual funds. Investors will be able to buy the targeted maturity and dollar amount over as long or as short a period of time as necessary, all with a high degree of liquidity.

Bond ETFs share the same benefits that equities enjoy in terms of lower costs, fees, and management control. But they differ in one major way. Studies have shown that over five-year periods during the past two decades only about a third of actively managed equity funds outpaced the benchmark S&P 500. However, 50 percent of active bond managers outperformed their benchmark index. Bond indexing may be less effective than equity indexing.

In general, passive bond investing, or indexing is attractive to the investor for several reasons. First, it offers low risk of underperforming the index. Second, the reduced investment management and custodial fees can enhance performance over time. Third, the investor does not have to worry about a manager's ability to forecast interest rates. This is important because changing interest rate levels, yield curve shapes, and yield spreads between bond sectors and individual bonds are all strategies based on the manager's ability to forecast interest rate movements. Finally, bond indexing allows the investor to know exactly what he or she is getting.

If the investor feels that the bond market is indeed an efficient one, and is a buy and hold investor, there is little need to look beyond the new bond ETFs to complete a portfolio asset allocation.

PROLIFERATION OF OTHER EQUITY PRODUCTS

These products do not qualify as basket securities under the definition used in this book, but are mentioned here as related developments.

Fidelity Investments is reintroducing stock baskets through its discount brokerage. Each basket will contain five to 50 stocks in a single industry sector or stock-picking approach. The initial stocks in each basket will be selected by Fidelity analysts or by outside providers such as Lehman Brothers. Fidelity says that investors will have some ability to modify the basket holdings.

The new offerings will allow individuals instead of professional fund managers to make their own decisions about when to sell individual stocks, giving them control over the timing of capital gains and losses. By contrast, mutual fund investors have no say in portfolio transactions even though they have to pay taxes on resulting distributions of capital gains.

Greater individual control is the major selling point for new types of personal funds that are starting to proliferate. Since mid-2000, FOLIOfn Inc. of Vienna, Virginia, has allowed investors to select which stocks they wish to hold in their personal fund-like portfolios, which they access through the company's website at *www.foliofn.com*.

The Investment Company Institute says the SEC should regulate these products as mutual funds. But sponsors, including Fidelity, say their equity products are not funds and need not be regulated as such.

At the time of this writing, Charles Schwab is marketing a pair of 10-stock portfolios in a Washington area pilot program. Schwab's U.S. Trust unit recommends the portfolio stocks. For an annual fee, investors receive periodic buy and sell recommendations for these aggressive growth and capital growth strategies and get to make some portfolio transactions without paying additional charges.

USE OF TRADING STRATEGIES

Going forward, more and more investors will turn to basket securities to implement entire investment strategies. For example, during the assembly or rebalancing of equity portfolios, the investor can equitize noninvested cash by using an S&P 500 Index basket allowing participation in any broad market rally. If investor analysis shows that utility share prices will weaken, the XLU SPDR can be shorted. If the portfolio already holds the S&P 500 SPDR, a short of XLU can bring the portfolio to a net neutral or net short position in the utility sector.

The use of basket securities, as discussed in Chapter 7, to implement strategies derived from economic analysis, should also increase in the future. For example, if capital spending or consumer demand is expected to decline with resultant lowered spending for technology issues, the technology SPDR (XLK) can be shorted. Further, if the investor feels that this reduced spending is a result of a peak in the business cycle and that assets will begin to move into the energy sector, he or she can short XLK and go long on the energy SPDR (XLE). The individual weights can be established according to the degree of confidence.

Baskets can also be used to exploit interest rate movements. If rates are rising, short the interest sensitive sector baskets such as the Utility SPDR (XLU), and buy the Financial SPDR (XLF). Higher rates force utility companies to pay higher dividends and force share prices down.

Baskets also facilitate hedging strategies, particularly those that protect against market reversals. As investors become more aware of this attribute, baskets will be used to a greater extent in this function. If, for example, the investor feels the economic outlook is attractive, purchasing a technology or cyclical basket and shorting a healthcare or consumer staples SPDR (XLP) can create a hedge. The former is closely tied to business cycle expansion, while the latter is too, but less so. Therefore, if the analysis is correct and the economy takes off, the investor gains on the long, and loses very little on the short. However, if the economy turns down, the short position will fall resulting in a profit that will offset the loss on the long side.

Basket securities are tremendously versatile and, as their numbers expand to include other sectors and the bond market, they have the capability to change the way money is managed. Similar to index funds but trading like stocks, they enjoy advantages from both worlds. They allow the investor to allocate, rebalance, hedge, or take concentrated market positions through a single security that can be bought and sold during market hours just like any other stock.

Section IV

Appendices

Appendix A — Fund Pages

TICKER	NAME	PAGE
	Small Cap	
IJR	iShares S&P Small Cap Index Fund	120
IJT	iShares S&P SmallCap/BARRA Growth Index Fund	121
IJS	iShares S&P 600/BARRA Value Index Fund	122
DSG	streetTRACKS Dow Jones U.S. Small Cap Growth Index Fund	123
DSV	streetTRACKS Dow Jones U.S. Small Cap Value Index Fund	124
	Mid Cap	
IJK	iShares S&P MidCap 400/Barra Growth	125
IJH	iShares S&P MidCap 400	126
MDY	SPDR Mid Cap 400	127
IJJ	iShares S&P MidCap 400/Barra Value	128
	Large Cap	
VTI	Vanguard Total Stock Market VIPERs	129
IWZ	iShares Russell 3000 Growth	130
ELG	streetTRACKS DJ US L/C Growth	131
IWF	iShares Russell 1000 Growth	132
IVW	iShares S&P 500/Barra Growth	133
SPY	SPDR 500	134
FFF	FORTUNE 500 INDEX FUND	135
IWV	iShares Russell 3000	136
IVV	iShares S&P 500/Index Fund	137
DGT	streetTRACKS Dow Jones Global Titans	138
IYY	streetTRACKS Dow Jones US Total Market	139
IWB	iShares Russell 1000	140
ELV	streetTRACKS DJ US L/C Value	141
DIA	Diamond Series Trust I	142
IVE	iShares S&P 500/Barra Value	143
IWW	iShares Russell 3000 Value	144
IWD	iShares Russell 1000 Value	145
IWO	iShares Rus 2000 Growth Index	146
IWM	iShares Russell 2000 Index	147
IWN	iShares Russell 2000 Value Index	148
	Financial	
XLF	SPDR Financial	149

TICKER	NAME	PAGE
IYF	iShares Dow Jone US Financial Sector	150
IYG	iShares Dow Jones US Financial Services	151

Health Care

IBB	iShares Nasdaq Biotechnology Index Fund	152
IYH	iShares Dow Jones US Healthcare	153

Basic Industry

IYM	iShares Dow Jones US Basic Materials	154
XLB	SPDR Basic Industries	155
XLI	SPDR Industrial	156
IYJ	iShares Dow Jones US Industrial	157

Utilities/Energy

XLE	SPDR Energy	158
IYE	iShares Dow Jones US Energy Sector	159
XLU	SPDR Utilities	160
IDU	iShares Dow Jones US Utilities	161

Chemicals

IYD	iShares Dow Jones US Chemicals	162

Consumer

XLV	SPDR Consumer Services	163
XLP	SPDR Consumer Staples	164

Technology

IYW	iShares Dow Jones US Technology	165
XLK	SPDR Technology	166
MTK	streetTRACKS MS HIGH TECH 35	167
IGM	iShares GS Technology Index Fund	168
FEF	FORTUNE e-50 INDEX FUND	169
QQQ	Nasdaq 100 Trust Series I	170
MII	streetTRACKS MS INTERNET INDEX	171
IYV	iShares Dow Jones US Internet	172
IYZ	iShares Dow Jones US Telecom	173

Cyclical/Noncyclical

XLY	SPDR Cyclical/Transportation	174
IYC	iShares Dow Jones US Consumer Cyclical	175
IYK	iShares Dow Jones US Consumer Noncyclical	176

Real Estate

ICF	iShares Cohen & Steers Realty	177
IYR	iShares Dow Jones US Real Estate	178
RWR	streetTRACKS Wilshire REIT Index Fund	179

Asia/Pacific Region

EWA	iShares MSCI Australia Index	180
EWT	iShares MSCI Taiwan Index	181

Appendix A — Fund Pages

TICKER	NAME	PAGE
EWS	iShares MSCI Singapore Index	182
EWH	iShares MSCI Hong Kong Index	183
EWY	iShars MSCI South Korea Index	184
EWJ	iShares MSCI Japan Index	185
EWM	iShares MSCI Malaysia Index	186
	Canada	
IKC	iShares S&P TSE 60 Index	187
EWC	iShares MSCI Canada Index	188
	Europe	
EWK	iShares MSCI Belgium Index	189
EWO	iShares MSCI Austria Index	190
EWP	iShares MSCI Spain Index	191
EWG	iShares MSCI Germany Index	192
EZU	iShares MSCI EMU Index	193
IEV	iShares MSCI Europe 350 Index Fund	194
EWI	iShares MSCI Italy Index	195
EWN	iShares MSCI Netherlands Index	196
EWL	iShares MSCI Switzerland Index	197
EWQ	iShares MSCI France Index	198
EWU	iShares MSCI UK Index	199
EWD	iShares MSCI Sweden Index	200
	Latin America	
EWZ	iShares MSCI Brazil Index	201
EWW	iShares MSCI Mexico Index	202

SMALL CAP

iShares S&P Small Cap 600 Index Fund (Symbol: IJR)

Objective

The S&P Small Cap 600 Index Fund seeks investment results that correspond generally to the price and yield performance, before fees and expenses, of the S&P SmallCap 600 Index (the "Index"). There is no assurance that the performance of the S&P SmallCap 600 Index can be fully matched.

Fund Details			
Options Traded:	No	Trading Increment:	0.01
Expense Ratio:	0.002	Min Trade Size:	1 Share
Ticker Symbol:	IJR	Marginable:	Yes
Short Selling Allowed:	Yes, uptick exempt		

Distribution History	2000	Quick Facts	
Ordinary Income	$0.45	Net Assets:	$216,300,000
Ordinary Income	$0.22	Shares Outstanding:	2,100,000
Short Term Capital Gains	$1.12	Dividend Yield:	0%
Long Term Capital Gains	$0.00	52 Week High:	$116.34
Return of Capital	$0.00	52 Week Low:	$93.80
Totals	$1.34		

Top Ten Holdings		Industry Groups	
1. Univ. Health Serv. Inc Class B	0.95%	1. Consumer Cyclicals	17.61%
2. Timberland Co	0.76%	2. Technology	16.28%
3. Cephalon Inc	0.75%	3. Financial	12.78%
4. Fidelity National Financial Inc	0.72%	4. Capital Goods	12.52%
5. Patterson Dental Co	0.65%	5. Health Care	11.93%
6. Eaton Vance Corp	0.65%	6. Consumer Staples	9.98%
7. Cullen/Frost Bankers Inc	0.62%	7. Energy	7.52%
8. Commerce Bancorp Inc NJ	0.61%	8. Utilities	4.38%
9. Varian Medical Systems Inc	0.61%	9. Basic Materials	3.91%
10. First American Financial Corp	0.60%	10. Transportation	2.73%
Total	6.92%	Total	99.67%

Trustee: Barclays Global Fund Advisors.
Administrator: Investors Bank & Trust Company.
Distributor: SEI Investments Distribution Co.

iShares S&P SmallCap 600/BARRA Growth Index Fund (Symbol: IJT)

Objective

iShares S&P SmallCap 600/BARRA Growth Index Fund seeks investment results that correspond generally to the price and yield performance, before fees and expenses, of the S&P SmallCap 600/BARRA Growth Index. There is no assurance that the performance of the S&P SmallCap 600/BARRA Growth Index can be fully matched.

Fund Details

Options Traded:	No	Trading Increment:	0.01
Expense Ratio:	0.0025	Min Trade Size:	1 Share
Ticker Symbol:	IJT	Marginable:	Yes
Short Selling Allowed:	Yes, uptick exempt		

Distribution History — 2000 | Quick Facts

Ordinary Income	$0.45	Net Assets:	$27,860,000
Ordinary Income	$0.00	Shares Outstanding:	400,000
Short Term Capital Gains	$0.62	Dividend Yield:	0%
Long Term Capital Gains	$0.00	52 Week High:	$90.00
Return of Capital	$0.00	52 Week Low:	$63.08
Totals	$0.62		

Top Ten Holdings | Industry Groups

1. Universal Health Services Inc Class B	2.05%	1. Health Care	20.59%
2. Timberland Co	1.63%	2. Technology	18.89%
3. Cephalon Inc	1.61%	3. Consumer Cyclicals	13.76%
4. Patterson Dental Co	1.40%	4. Capital Goods	12.27%
5. Eaton Vance Corp	1.40%	5. Consumer Staples	11.99%
6. Cullen/Frost Bankers Inc	1.33%	6. Energy	8.97%
7. Commerce Bancorp Inc NJ	1.32%	7. Financial	7.70%
8. Varian Medical Systems Inc	1.32%	8. Basic Materials	2.83%
9. RSA Security Inc	1.28%	9. Transportation	2.65%
10. Shaw Group Inc	1.24%	10. Communication Services	0.29%
Total	14.58%	Total	99.93%

Trustee: Barclays Global Fund Advisors.
Administrator: Investors Bank & Trust Company.
Distributor: SEI Investments Distribution Co.

iShares S&P SmallCap 600/BARRA Value Index Fund (Symbol: IJS)

Objective

iShares S&P SmallCap 600/BARRA Value Index Fund seeks investment results that correspond generally to the price and yield performance, before fees and expenses, of the S&P SmallCap 600/BARRA Value Index. There is no assurance that the performance of the S&P SmallCap 600/BARRA Value Index can be fully matched.

Fund Details

Options Traded:	No	Trading Increment:	0.01
Expense Ratio:	0.0025	Min Trade Size:	1 Share
Ticker Symbol:	IJS	Marginable:	Yes
Short Selling Allowed:	Yes, uptick exempt		

Distribution History 2000

Ordinary Income	$0.45
Ordinary Income	$0.26
Short Term Capital Gains	$0.34
Long Term Capital Gains	$0.00
Return of Capital	$0.00
Totals	$0.60

Quick Facts

Net Assets:	$60,712,000
Shares Outstanding:	800,000
Dividend Yield:	1%
52 Week High:	$85.62
52 Week Low:	$35.00

Top Ten Holdings

1. Fidelity National Financial Inc	1.34%
2. First American Financial Corp	1.11%
3. Centura Banks Inc	1.01%
4. Alpharma Inc	0.94%
5. Smithfield Foods Inc	0.88%
6. Pride International Inc	0.88%
7. DR Horton Inc	0.88%
8. Raymond James Financial Corp	0.86%
9. Coventry Health Care Inc	0.84%
10. Downey Financial Corp	0.83%
Total	9.57%

Industry Groups

1. Consumer Cyclicals	20.94%
2. Financial	17.19%
3. Technology	14.04%
4. Capital Goods	12.72%
5. Consumer Staples	8.24%
6. Utilities	8.18%
7. Energy	6.27%
8. Basic Materials	4.85%
9. Health Care	4.44%
10. Transportation	2.81%
Total	99.67%

Trustee: Barclays Global Fund Advisors.
Administrator: Investors Bank & Trust Company.
Distributor: SEI Investments Distribution Co.

streetTRACKS Dow Jones U.S. Small Cap Growth Index Fund (Symbol: DSG)

Objective

The streetTRACKS Dow Jones U.S. Small Cap Growth Index Fund seeks investment results that correspond generally to the price and yield performance, before fees and expenses, of the Dow Jones U.S. Small Cap Growth Index. There is no assurance that the performance of the Dow Jones U.S. Small Cap Growth Index can be fully matched.

Fund Details			
Options Traded:	No	Tracking Increment:	0.01
Expense Ratio:	0.0025	Min Trade Size:	1 Share
Ticker Symbol:	DSG	Marginable:	Yes
Short Selling Allowed:	Yes, uptick exempt		

Distribution History	2000	Quick Facts	
Ordinary Income	$0.45	Net Assets:	$13,540,000
Ordinary Income	$0.00	Shares Outstanding:	200,000
Short Term Capital Gains	$0.00	Dividend Yield:	0%
Long Term Capital Gains	$0.00	52 Week High:	$100.94
Return of Capital	$0.00	52 Week Low:	$44.50
Total	$0.00		

Top Ten Holdings		Industry Groups	
1. SEI Investments Co.	1.39%	1. Healthcare	22.28%
2. International Game Techn	1.12%	2. Technology	20.17%
3. Netiq	1.11%	3. Industrial	16.55%
4. Macrovision Corp.	1.09%	4. Consumer Cyclicals	15.63%
5. Lincare Hldgs Inc.	0.96%	5. Energy	7.46%
6. Invitrogen Corp.	0.94%	6. Financials	7.17%
7. Trigon Healthcare Inc.	0.94%	7. Consumer Non-Cyclical	5.60%
8. Universal Health Svcs Inc.	0.90%	8. Telecommunications	3.03%
9. Expeditores Int'l Wash Inc.	0.89%	9. Basic Materials	1.25%
10. US Airways Group Inc.	0.88%	10. Utilities	0.86%
Total	10.21%	Total	100.00%

Trustee: State Street Global Advisors
Administrator: State Street Bank & Trust Co.
Distributor: State Street Capital Markets, LLC

streetTRACKS Dow Jones U.S. Small Cap Value Index Fund (Symbol: DSV)

Objective

The streetTRACKS Dow Jones U.S. Small Cap Value Index Fund seeks investment results that correspond generally to the price and yield performance, before fees and expenses, of the Dow Jones U.S. Small Cap Value Index. There is no assurance that the performance of the Dow Jones U.S. Small Cap Value Index can be fully matched.

Fund Details			
Options Traded:	No	Tracking Increment:	0.01
Expense Ratio:	0.0025	Min Trade Size:	1 Share
Ticker Symbol:	DSV	Marginable:	Yes
Short Selling Allowed:	Yes, uptick exempt		
Distribution History	**2000**	**Quick Facts**	
Ordinary Income	$0.45	Net Assets:	$11,712,000
Ordinary Income	$0.93	Shares Outstanding:	100,000
Short Term Capital Gains	$0.00	Dividend Yield:	0%
Long Term Capital Gains	$0.00	52 Week High:	$124.30
Return of Capital	$0.00	52 Week Low:	$53.50
Total	$0.93		
Top Ten Holdings		**Industry Groups**	
1. Laboratory Corp. Amer. Hld.	1.49%	1. Financials	42.78%
2. Old Republic International Corp.	1.28%	2. Industrial	17.04%
3. Everest Reinsurance Group	1.20%	3. Consumer Cyclical	13.10%
4. Radian Group Inc.	1.04%	4. Utilities	8.53%
5. Astoria Fin'l Corp	0.99%	5. Basic Materials	5.20%
6. Litton Industries Inc.	0.94%	6. Consumer Non-Cyclicals	4.71%
7. Humana Inc.	0.88%	7. Healthcare	4.50%
8. Questar Corp.	0.81%	8. Technology	3.06%
9. Bank Utd. Corp.	0.81%	9. Energy	0.67%
10. FirstMerit Corp.	0.79%	10. Telecommunications	0.40%
Total	10.21%	Total	100.00%

Trustee: State Street Global Advisors
Administrator: State Street Bank & Trust Co.
Distributor: State Street Capital Markets, LLC

Appendix A — Fund Pages

MID CAP BASKETS

iShares S&P MidCap 400/BARRA Growth Index Fund
(Symbol: IJK)

Objective

iShares S&P MidCap 400/BARRA Growth Index Fund seeks investment results that correspond generally to the price and yield performance, before fees and expenses, of the S&P MidCap 400/BARRA Growth Index. There is no assurance that the performance of the S&P MidCap 400/BARRA Growth Index can be fully matched.

Fund Details			
Options Traded:	No	Trading Increment:	0.01
Expense Ratio:	0.0025	Min Trade Size:	1 Share
Ticker Symbol:	IJK	Marginable:	Yes
Short Selling Allowed:	Yes, uptick exempt		

Distribution History	2000	Quick Facts	
Ordinary Income	$0.45	Net Assets:	$165,632,000
Ordinary Income	$0.00	Shares Outstanding:	1,600,000
Short Term Capital Gains	$0.47	Dividend Yield:	0%
Long Term Capital Gains	$0.00	52 Week High:	$151.50
Return of Capital	$0.00	52 Week Low:	$93.24
Totals	$0.47		

Top Ten Holdings		Industry Groups	
1. Millennium Pharmaceuticals Inc	2.94%	1. Technology	31.80%
2. Waters Corp Corp	2.41%	2. Health Care	21.75%
3. Concord EFS Inc	2.12%	3. Consumer Cyclicals	16.13%
4. Idec Pharmaceuticals Corp	2.03%	4. Financial	8.29%
5. Cintas Corp	2.01%	5. Consumer Staples	6.28%
6. Genzyme Corp - General Division	1.91%	6. Energy	5.85%
7. Univision Communications Inc Class A	1.89%	7. Capital Goods	5.31%
8. DST Systems Inc	1.87%	8. Transportation	1.46%
9. Rational Software Inc	1.64%	9. Utilities	1.18%
10. Cadence Design systems Inc	1.51%	10. Communications Services	1.04%
Total	17.39%	Total	99.09%

Trustee: Barclays Global Fund Advisors.
Administrator: Investors Bank & Trust Company.
Distributor: SEI Investments Distribution Co.

iShares S&P Midcap 400 Index Fund (Symbol: IJH)

Objective

The S&P MidCap 400 Index Fund seeks investment results that correspond generally to the price and yield performance, before fees and expenses, of the S&P MidCap 400 IndexTM. There is no assurance that the performance of the S&P MidCap 400 Index can be fully matched.

Fund Details			
Options Traded:	No	Trading Increment:	0.01
Expense Ratio:	0.002	Min Trade Size:	1 Share
Ticker Symbol:	IJH	Marginable:	Yes
Short Selling Allowed:	Yes, uptick exempt		

Distribution History	2000	Quick Facts	
Ordinary Income	$0.45	Net Assets:	$162,977,500
Ordinary Income	$0.47	Shares Outstanding:	1,750,000
Short Term Capital Gains	$0.30	Dividend Yield:	1%
Long Term Capital Gains	$0.00	52 Week High:	$110.17
Return of Capital	$0.00	52 Week Low:	$86.32
Totals	$0.77		

Top Ten Holdings		Industry Groups	
1. Millennium Pharmaceuticals Inc	1.47%	1. Technology	20.72%
2. Waters Corp Corp	1.21%	2. Financial	15.30%
3. Concord EFS Inc	1.06%	3. Consumer Cyclicals	14.03%
4. Idec Pharmaceuticals Corp	1.01%	4. Health Care	12.82%
5. Cintas Corp	1.00%	5. Utilities	7.84%
6. Genzyme Corp - General Division	0.96%	6. Consumer Staples	7.82%
7. Univision Communications Inc Class A	0.95%	7. Energy	7.25%
8. DST Systems Inc	0.94%	8. Capital Goods	6.81%
9. Rational Software Inc	0.82%	9. Basic Materials	3.73%
10. Cadence Design systems Inc	0.76%	10. Transportation	1.96%
Total	8.71%	Total	98.28%

Trustee: Barclays Global Fund Advisors.
Administrator: Investors Bank & Trust Company.
Distributor: SEI Investments Distribution Co.

Appendix A — Fund Pages

STANDARD & POOR'S MIDCAP 400 DEPOSITARY RECEIPTS (SYMBOL: MDY)

Objective

The MidCap SPDR Trust Series I is a pooled investment designed to provide investment results that generally correspond to the price and yield performance of the S&P 400™ Index. There is no assurance that the performance of the S&P 400 Index can be fully matched.

Fund Details

Options Traded:	Yes	Trading Increment:	0.01
Expense Ratio:	0.0025	Min Trade Size:	1 Share
Ticker Symbol:	MDY	Marginable:	Yes
Short Selling Allowed:	Yes, uptick exempt		

Distribution History	2000	Quick Facts	
Ordinary Income	$0.78	Net Assets:	$3,658,952,880
Short Term Capital Gains	$0.00	Shares Outstanding:	41,999,000
Long Term Capital Gains	$0.00	Dividend Yield:	1%
Return of Capital	$0.00	52 Week High:	$101.12
Totals	$0.78	52 Week Low:	$77.69

Top Ten Holdings		Industry Groups	
1. Millennium Pharmaceuticals	1.47%	1. Technology	20.56%
2. Waters Corporation	1.21%	2. Financials	15.30%
3. Concord EFS Inc.	1.06%	3. Consumer Cyclicals	13.76%
4. IDEC Pharmaceuticals	1.01%	4. Health Care	12.83%
5. Cintas Corporation	1.00%	5. Consumer Staples	7.82%
6. Genzyme Corp.	0.96%	6. Utilities	7.59%
7. Univision Communications	0.95%	7. Energy	7.26%
8. DST Systems Inc.	0.94%	8. Capital Goods	6.59%
9. Rational Software	0.82%	9 Basic Materials	3.72%
10. Cadence Design Systems	0.76%	10. Transportation	1.95%
Total	10.19%	Total	97.39%

Trustee: Bank of New York
Administrator: PDR Services LLC
Distributor: ALPS Mutual Funds Inc.

iShares S&P MidCap 400/Barra Value Index Fund (Symbol: IJJ)

Objective

iShares S&P MidCap 400/BARRA Value Index Fund seeks investment results that correspond generally to the price and yield performance, before fees and expenses, of the S&P MidCap 400/BARRA Value Index. There is no assurance that the performance of the S&P MidCap 400/BARRA Value Index can be fully matched.

Annual Performance at NAV

	Inception	2000	2001
IJJ	7/24/2000	18.89%	-3.51%
S&P MC BARValue Index		19.05%	-3.45%

Distribution History

	2000	2001
Ordinary Income	$0.51	$0.24
Short Term Capital Gains	$0.14	$0.00
Long Term Capital Gains	$0.00	$0.00
Return of Capital	$0.00	$0.00
Totals	$0.65	$0.24

Fund Details

Expense Ratio:	0.25%
Ticker Symbol:	IJJ
Trading Increment:	$0.01
Min. Trade Size:	1 Share
Marginable:	Yes
Options Traded:	No
Short Selling Allowed:	Yes, uptick exempt

Quick Facts

Net Assets:	$8,091,000
Shares Outstanding:	90,000
Dividend Yield:	1%
52 Week High:	$95.12
52 Week Low:	$40
Information as of:	7/17/01

Top Ten Holdings

1. M & T Bank Corp	1.58%
2. RJ Reynolds Tob Hldgs Inc	1.34%
3. Telephone & Data Sys. Inc	1.29%
4. Marshall & Ilsley Corp	1.27%
5. Weatherford Int'l Inc	1.27%
6. Ensco International Inc	1.14%
7. Atmel Corp	1.07%
8. Zions Bancorp	1.06%
9. Jones Apparel Group Inc	1.06%
10. North Fork BanCorp Inc	1.00%
Total	12.08%

Economic Sectors

1. Financial	22.58%
2. Utilities	14.22%
3. Consumer Cyclicals	12.62%
4. Consumer Staples	10.13%
5. Technology	9.34%
6. Capital Goods	8.37%
7. Energy	7.76%
8. Basic Materials	6.74%
9. Health Care	3.62%
10. Transportation	2.33%
Total	97.71%

Advisor: Barclays Global Fund Advisors
Administrator: Investors Bank & Trust
Distributor: SEI Investments Dist. Co.

Appendix A — Fund Pages

LARGE CAP BASKETS
VANGUARD TOTAL STOCK MARKET VIPERs (SYMBOL: VTI)

Objective

The Vanguard Total Stock Market VIPERs are designed to provide investors with performance, before fees and expenses, that generally corresponds to the performance of the Wilshire 5000 Total Market Index. There can be no assurance that the returns of the Wilshire 5000 Total Market Index can be fully matched. The Fund invests all or substantially all of its assets in a representative sample of the stocks that make up the index.

Annual Performance at NAV

	Inception	2001
VTI	5/31/2001	
Wilshire 5000 Total Mkt Idx		

Distribution History

	2001
Ordinary Income	$0.00
Short Term Capital Gains	$0.00
Long Term Capital Gains	$0.00
Return of Capital	$0.00
Totals	$0.00

Fund Details

Expense Ratio:	0.15%
Ticker Symbol:	VTI
Trading Increment:	$0.01
Min. Trade Size:	1 Share
Marginable:	Yes
Options Traded:	Yes
Short Selling Allowed: Yes, uptick exempt	

Quick Facts

Net Assets:	
Shares Outstanding:	
Dividend Yield:	1.00%
52 Week High:	$118
52 Week Low:	$107.26

Information as of: 7/17/01

Top Ten Holdings

1. General Electric Co.	3.22%
2. Microsoft Corp.	2.40%
3. Exxon Mobil Corp.	2.37%
4. Pfizer, Inc.	2.20%
5. Wal-Mart Stores, Inc.	1.93%
6. American Intern'l Group	1.78%
7. Citigroup, Inc.	1.77%
8. Johnson & Johnson.	1.46%
9. IBM	1.40%
10. SBC Communications	1.38%
Total	19.91%

Economic Sectors

1. Financial Services	19.90%
2. Technology	16.50%
3. Consumer Discret. & Svcs.	13.60%
4. Health Care	13.50%
5. Utilities	9.50%
6. Consumer Staples	6.40%
7. Other	5.30%
8. Integrated Oils	3.60%
9. Producer Durables	3.40%
10. Other Energy	3.20%
Total	94.90%

Advisor: The Vanguard Group
Administrator: The Vanguard Group
Distributor: Vanguard Marketing Corp.

iShares Russell 3000 Growth Index Fund (Symbol: IWZ)

Objective

iShares Russell 3000 Growth Index Fund seeks investment results that correspond generally to the price and yield performance, before fees and expenses, of the Russell 3000 Growth Index. There is no assurance that the performance of the Russell 3000 Growth Index can be fully matched.

Fund Details

Options Traded:	No	Trading Increment:	0.01
Expense Ratio:	0.0025	Min Trade Size:	1 Share
Ticker Symbol:	IWZ	Marginable:	Yes
Short Selling Allowed:	Yes, uptick exempt		

Distribution History

	2000
Ordinary Income	$0.03
Short Term Capital Gains	$0.00
Long Term Capital Gains	$0.00
Return of Capital	$0.00
Total	$0.03

Quick Facts

Net Assets:	$21,020,000
Shares Outstanding:	500,000
Dividend Yield:	0%
52 Week High:	$71.97
52 Week Low:	$25.00

Top Ten Holdings

1. General Elec. Co	7.58%
2. Pfizer Inc	5.09%
3. Cisco Sys Inc	4.80%
4. Intel Corp	3.53%
5. Microsoft Corp	3.03%
6. EMC Corp	2.54%
7. Wal-Mart Stores Inc.	2.30%
8. Oracle Corp	2.17%
9. Int'l Business Machines	1.92%
10. Merck & Co. Inc.	1.90%
Total	34.86%

Industry Groups

1. Technology	40.52%
2. Health Care	21.96%
3. Consumer Discretionary	12.80%
4. Other	7.68%
5. Consumer Staples	4.22%
6. Financial Services	4.04%
7. Producer Durables	2.78%
8. Utilities	2.60%
9. Other Energy	2.38%
10. Materials & Processing	0.49%
Total	99.49%

Trustee: Barclays Global Fund Advisors.
Administrator: Investors Bank & Trust Company.
Distributor: SEI Investments Distribution Co.

Appendix A — Fund Pages

streetTRACKS Dow Jones U.S. Large Cap Growth Index Fund (Symbol: ELG)

Objective

The streetTRACKS Dow Jones U.S. Large Cap Growth Index Fund seeks investment results that correspond generally to the price and yield performance, before fees and expenses, of the Dow Jones U.S. Large Cap Growth Index. There is no assurance that the performance of the Dow Jones U.S. Large Cap Growth Index can be fully matched.

Fund Details

Options Traded:	No	Tracking Increment:	0.01
Expense Ratio:	0.002	Min Trade Size:	1 Share
Ticker Symbol:	ELG	Marginable:	Yes
Short Selling Allowed:	Yes, uptick exempt		

Distribution History 2000

Ordinary Income	$0.00
Ordinary Income	$0.02
Short Term Capital Gains	$0.00
Long Term Capital Gains	$0.00
Return of Capital	$0.00
Total	$0.02

Quick Facts

Net Assets:	$23,068,000
Shares Outstanding:	400,000
Dividend Yield:	0%
52 Week High:	$97.31
52 Week Low:	$40.00

Top Ten Holdings

1. General Electric Co.	11.82%
2. Pfizer, Inc.	7.20%
3. Cisco Systems, Inc.	6.68%
4. Microsoft Corp.	4.90%
5. Intel Corp.	4.74%
6. Wal-Mart, Inc.	3.66%
7. EMC Corp.	3.61%
8. Coca Cola Co.	3.23%
9. Oracle Corp.	3.10%
10. Home Depot Inc.	2.63%
Total	51.57%

Industry Groups

1. Technology	43.76%
2. Industrial	18.27%
3. Consumer Cyclicals	15.38%
4. Healthcare	12.00%
5. Consumer Non-Cyclicals	5.92%
6. Telecommunications	2.50%
7. Financials	1.21%
8. Utilities	0.52%
9. Energy	0.44%
Total	100%

Advisor: State Street Global Advisors
Administrator: State Street Bank & Trust Co.
Distributor: State Street Capital Markets, LLC

iShares Russell 1000 Growth Index Fund (Symbol: IWF)

Objective

The Russell 1000 Growth Index Fund seeks investment returns that correspond generally to the price and yield performance, before fees and expenses, of the Russell 1000 Growth Index. There is no assurance that the performance of the Russell 1000 Growth Index can be fully matched.

Fund Details			
Options Traded:	No	Trading Increment:	$0.01
Expense Ratio:	0.20%	Min Trade Size:	1 Share
Ticker Symbol:	IWF	Marginable:	Yes
Short Selling Allowed:	Yes, uptick exempt		

Distribution History	2000	Quick Facts	
		Net Assets:	$166,729,500
Ordinary Income	$0.09	Shares Outstanding:	3,150,000
Short Term Capital Gains	$0.11	Dividend Yield:	0%
Long Term Capital Gains	$0.00	52 Week High:	$92.33
Return of Capital	$0.00	52 Week Low:	$46.66
Total	$0.20	As of Apr. 16, 2001	

Top Ten Holdings		Industry Groups	
1. General Electric Co	7.72%	1. Technology	50.34%
2. Cisco Systems Inc	5.82%	2. Health Care	16.20%
3. Pfizer Inc	4.21%	3. Consumer Discretionary	12.09%
4. Intel Corp.	4.14%	4. Other	7.77%
5. Microsoft Corp	3.50%	5. Consumer Staples	3.10%
6. EMC Corp	3.21%	6. Financial Services	3.09%
7. Sun Microsystems Inc	2.75%	7. Utilities	2.67%
8. Oracle Corp	2.52%	8. Producer Durables	2.19%
9. Intl Business Machines	2.17%	9. Other Energy	2.01%
10. America Online Inc	1.84%	10. Auto & Transportation	0.34%
Total	37.89%	Total	99.80%

Trustee: Barclays Global Fund Advisors.
Administrator: Investors Bank & Trust Company.
Distributor: SEI Investments Distribution Co.

Appendix A — Fund Pages

iShares S&P 500/BARRA Growth Index Fund (Symbol: IVW)

Objective

The S&P 500/BARRA Growth Index Fund seeks investment results that correspond generally to the price and yield performance, before fees and expenses, of the S&P 500/BARRA Growth Index. There is no assurance that the performance of the S&P 500/BARRA Growth Index can be fully matched.

Fund Details			
Options Traded:	No	Trading Increment:	0.01
Expense Ratio:	0.0018	Min Trade Size:	1 Share
Ticker Symbol:	IVW	Marginable:	Yes
Short Selling Allowed:	Yes, uptick exempt		

Distribution History	2000	Quick Facts	
Ordinary Income	$0.45	Net Assets:	$142,750,000
Ordinary Income	$0.16	Shares Outstanding:	2,500,000
Short Term Capital Gains	$0.11	Dividend Yield:	0%
Long Term Capital Gains	$0.00	52 Week High:	$94.25
Return of Capital	$0.00	52 Week Low:	$52.88
Totals	$0.27		

Top Ten Holdings		Industry Groups	
1. General Electric	8.28%	1. Technology	32.68%
2. Pfizer Inc	5.06%	2. Health Care	26.19%
3. Cisco Systems Inc	4.80%	3. Consumer Staples	14.13%
4. Wal-Mart Stores	4.14%	4. Financial	8.83%
5. Microsoft Corp	4.03%	5. Capital Goods	8.79%
6. American International Group Inc	4.00%	6. Consumer Cyclicals	8.14%
7. Merck & Co Inc	3.77%	7. Communications Services	0.66%
8. Intel Corp	3.53%	8. Utilities	0.39%
9. Oracle Systems Corp	2.84%	9. Basic Materials	0.10%
10. The Coca-Cola Co	2.64%	10. Transportation	5.00%
Total	43.09%	Total	99.96%

Trustee: Barclays Global Fund Advisors.
Administrator: Investors Bank & Trust Company.
Distributor: SEI Investments Distribution Co.

STANDARD & POOR'S DEPOSITARY RECEIPTS
(SYMBOL: SPY)

Objective

The SPDR Trust Series I is a pooled investment designed to provide investment results that generally correspond to the price and yield performance, before fees and expenses, of the S&P 500 Index. There is no assurance that the performance of the S&P 500 Index can be fully matched.

Annual Performance

	Inception	1993	1994	1995	1996	1997	1998	1999	2000
SPY	01/29/1993	8.92%	1.16%	37.22%	22.70%	33.06%	28.35%	20.86%	-9.14%
S&P 500 Index		9.17%	1.32%	37.58%	22.96%	33.36%	28.58%	21.04%	-9.11%

Distribution History

	1993	1994	1995	1996	1997	1998	1999	2000
Ordinary Income	$1.13	$1.13	$1.30	$1.38	$1.38	$1.42	$1.44	$1.51
Short Term Capital Gain	$0.00	$0.00	$0.00	$0.00	$0.00	$0.00	$0.00	$0.00
Long Term Capital Gain	$0.00	$0.00	$0.00	$0.09	$0.00	$0.00	$0.00	$0.00
Return of Capital	$0.00	$0.00	$0.00	$0.00	$0.00	$0.00	$0.00	$0.00
Totals	$1.13	$1.13	$1.30	$1.47	$1.38	$1.42	$1.44	$1.51

Trust Details

Options Traded:	No
Expense Ratio:	0.12%
Ticker Symbol:	SPY
Trading Increment:	$0.01
Min Trade Size:	1 Share
Marginable:	Yes
Short Selling Allowed:	Yes, uptick exempt

Quick Facts   As of Apr. 18, 2001

Net Assets:	$26,067,373,760
Shares Outstanding:	218,576,000
Dividend Yield:	1%
52 Week High:	$153.59
52 Week Low:	$108.04

Top Ten Holdings

1. General Electric Co.	4.03%
2. Exxon Mobil Corp.	2.57%
3. Pfizer, Inc.	2.47%
4. Cisco Systems	2.34%
5. Citigroup Inc.	2.18%
6. Wal Mart Stores Inc.	2.02%
7. Microsoft Corp	1.97%
8. American Intl. Group Inc.	1.95%
9. Merck & Co., Inc.	1.83%
10. Intel Corp.	1.72%
Total	23.08%

Economic Sectors

1. Technology	21.81%
2. Financials	17.03%
3. Health Care	13.94%
4. Consumer Staples	10.66%
5. Capital Goods	8.92%
6. Consumer Cyclicals	7.56%
7. Energy	6.24%
8. Communication Services	5.44%
9. Utilities	3.93%
10. Basic Materials	2.39%
Total	97.92%

Trustee: State Street Bank
Sponsor: PDR Services LLC
Distributor: ALPS Mutual Funds Inc.

Appendix A — Fund Pages
135

FORTUNE 500 INDEX TRACKING STOCK
(SYMBOL: FFF)

Objective

The FORTUNE 500 Index Tracking Stock seeks investment results that correspond to the price and yield performance, before fees and expenses, of the FORTUNE 500 Index. There is no assurance that the performance of the FORTUNE 500 Index can be fully matched.

Fund Details			
Options Traded:	Yes	Tracking Increment:	0.01
Expense Ratio:	0.002	Min Trade Size:	1 Share
Ticker Symbol:	FFF	Marginable:	Yes
Short Selling Allowed:	Yes, uptick exempt		

Distribution History	2000	Quick Facts	
Ordinary Income	$0.00	Net Assets:	$50,820,000
Ordinary Income	$0.00	Shares Outstanding:	600,000
Short Term Capital Gains	$0.00	Dividend Yield:	0%
Long Term Capital Gains	$0.00	52 Week High:	$98.91
Return of Capital	$0.00	52 Week Low:	$77.05
Totals	$0.00		

Top Ten Holdings		Industry Groups	
1. General Electric	4.53%	1. Financials	24.09%
2. Exxon Mobil Corporation	2.88%	2. Technology	18.05%
3. Pfizer Inc	2.77%	3. Health Care	14.17%
4. Cisco Systems	2.58%	4. Consumer Staples	12.52%
5. Citigroup Inc	2.45%	5. Consumer Cyclicals	7.89%
6. Wall-Mart Stores	2.26%	6. Energy	5.94%
7. Microsoft Corp	2.21%	7. Communication Services	5.93%
8. American Int'l Group Inc.	2.19%	8. Capital Goods	4.67%
9. Merck & Co. Inc.	2.06%	9. Utilities	3.04%
10. Intel Corp	1.93%	10. Basic Materials	2.18%
Total	25.86%	Total	98.48%

Trustee: State Street Global Advisors
Administrator: State Street Bank & Trust Co.
Distributor: State Street Capital Markets, LLC

iShares Russell 3000 Index Fund (Symbol: IWV)

Objective

The Russell 3000 Index Fund seeks investment results that correspond generally to the price and yield performance, before fees and expenses, of the Russell 3000 Index. There is no assurance that the performance of the Russell 3000 Index can be fully matched.

Fund Details			
Options Traded:	No	Trading Increment:	0.01
Expense Ratio:	0.002	Min Trade Size:	1 Share
Ticker Symbol:	IWV	Marginable:	Yes
Short Selling Allowed:	Yes, uptick exempt		

Distribution History	2000	Quick Facts	
Ordinary Income	$0.45	Net Assets:	$393,450,000
Short Term Capital Gains	$0.02	Shares Outstanding:	6,100,000
Long Term Capital Gains	$0.00	Dividend Yield:	1%
Return of Capital	$0.00	52 Week High:	$84.69
Total	$0.36	52 Week Low:	$59.44

Top Ten Holdings		Industry Groups	
1. General Electric Co.	3.70%	1. Technology	19.97%
2. Exxon Mobil Corp.	2.36%	2. Financial Services	19.79%
3. Pfizer Inc	2.26%	3. Health Care	14.78%
4. Cisco Systems Inc	2.14%	4. Consumer Discretionary	10.91%
5. Citigroup Inc.	2.01%	5. Utilities	8.81%
6. Merck & Co.	1.68%	6. Consumer Staples	6.31%
7. Intel Corp.	1.57%	7. Other	4.92%
8. American International Group Inc	1.54%	8. Integrated Oils	3.75%
9. Microsoft Corp.	1.35%	9. Producer Durables	3.38%
10. SBC Communications Inc.	1.26%	10. Materials & Processing	2.84%
Total	19.87%	Total	95.46%

Trustee: Barclays Global Fund Advisors.
Administrator: Investors Bank & Trust Company.
Distributor: SEI Investments Distribution Co.

iShares S&P 500 Index Fund
(Symbol: IVV)

Objective

iShares S&P 500 Index Fund seeks investment results that correspond generally to the price and yield performance, before fees and expenses, of the S&P 500 Index. There is no assurance that the performance of the S&P 500 Index can be fully matched.

Fund Details

Options Traded:	No	Trading Increment:	0.01
Expense Ratio:	0.0009	Min Trade Size:	1 Share
Ticker Symbol:	IVV	Marginable:	Yes
Short Selling Allowed:	Yes, uptick exempt		

Distribution History 2000

		Quick Facts	
Ordinary Income	$0.45	Net Assets:	$2,940,357,500
Ordinary Income	$0.77	Shares Outstanding:	24,950,000
Short Term Capital Gains	$0.07	Dividend Yield:	1%
Long Term Capital Gains	$0.00	52 Week High:	$153.47
Return of Capital	$0.00	52 Week Low:	$108.38
Totals	$0.84		

Top Ten Holdings

		Industry Groups	
1. General Electric	4.05%	1. Technology	21.83%
2. Exxon Mobil Corp	2.58%	2. Financial	17.33%
3. Pfizer Inc	2.48%	3. Health Care	14.02%
4. Cisco Systems Inc	2.35%	4. Consumer Staples	11.34%
5. Citigroup Inc	2.19%	5. Capital Goods	8.95%
6. Wal-Mart Stores Inc	2.02%	6. Consumer Cyclicals	7.57%
7. Microsoft Corp	1.97%	7. Energy	6.41%
8. American International Group Inc	1.96%	8. Communications Services	5.46%
9. Merck & Co Inc	1.84%	9. Utilities	3.94%
10. Intel Corp	1.73%	10. Basic Materials	2.39%
Total	23.17%	Total	99.24%

Trustee: Barclays Global Fund Advisors.
Administrator: Investors Bank & Trust Company.
Distributor: SEI Investments Distribution Co.

streetTRACKS Dow Jones Global Titans Index Fund (Symbol: DGT)

Objective

The streetTRACKS Dow Jones Global Titans Index Fund seeks investment results that correspond to the price and yield performance, before fees and expenses, of the Dow Jones Global Titans Index. There is no assurance that the performance of the Dow Jones Global Titans Index can be fully matched.

Fund Details		Quick Facts	
Options Traded:	No	Tracking Increment:	0.01
Expense Ratio:	0.005	Min Trade Size:	1 Share
Ticker Symbol:	DGT	Marginable:	Yes
Short Selling Allowed:	Yes, uptick exempt		

Distribution History	2000	Quick Facts	
Ordinary Income	$0.00	Net Assets:	$21,093,000
Ordinary Income	$0.13	Shares Outstanding:	300,000
Short-Term Capital Gains	$0.00	Dividend Yield:	0%
Long Term Capital Gains	$0.00	52 Week High:	$85.50
Return of Capital	$0.00	52 Week Low:	$64.37
Total	$0.13		

Top Ten Holdings		Industry Groups	
1. General Electric Co.	11.82%	1. Technology	43.76%
1. General Electric	8.10%	1. Financials	20.38%
2. Exxon Mobil	5.15%	2. Technology	20.21%
3. Cisco Systems	4.58%	3. Telecommunications	14.20%
4. Citigroup	4.38%	4. Energy	12.39%
5. Vodafone Group PLC	3.85%	5. Healthcare	9.25%
6. Merch & Co.	3.68%	6. Consumer Cyclicals	8.25%
7. Nokia	3.58%	7. Industrials	8.10%
8. Microsoft Corp.	3.36%	8. Consumer Non-Cyclical	7.22%
9. American Int'l Group Inc.	3.36%		
10. Intel Corp.	3.25%		
Total	43.27%	Total	100%

Advisor: State Street Global Advisors
Administrator: State Street Bank & Trust Co.
Distributor: State Street Capital Markets, LLC

iShares Dow Jones U.S. Total Market Index Fund
Symbol: IYY

Objective

iShares Dow Jones U.S. Total Market Index Fund seeks investment results that correspond generally to the price and yield performance, before fees and expenses, of the Dow Jones U.S. Total Market Index. There is no assurance that the performance of the Dow Jones U.S. Total Market Index can be fully matched.

Fund Details

Options Traded:	No	Trading Increment:	$0.01
Expense Ratio:	0.20%	Min Trade Size:	1 Share
Ticker Symbol:	IYY	Marginable:	Yes
Short Selling Allowed:	Yes, Uptick exempt		

Distribution History 2000

Ordinary Income	$0.24
Short Term Capital Gains	$0.01
Long Term Capital Gains	$0.00
Return of Capital	$0.00
Total	$0.25

(Dividends paid quarterly and capital gains at least annually)

Quick Facts

Net Assets:	$62,744,000
Shares Outstanding:	1,150,000
Dividend Yield:	1%
52 Week High:	$71.97
52 Week Low:	$50

As of Apr. 16, 2001

Top Ten Holdings

General Electric Co	4.03%
Cisco Systems Inc	2.74%
Exxon Mobil Corp	2.18%
Pfizer Inc	2.00%
Intel Corp	1.86%
Microsoft Corp	1.78%
Citigroup Inc.	1.72%
EMC Corp.	1.52%
Intl'. Business Machines Co	1.40%
American International Group Inc	1.34%
Total	20.56%

Industry Groups

Technology	26.48%
Financial	15.81%
Industrial	12.82%
Healthcare	11.81%
Consumer, Cyclical	10.39%
Consumer, Non-Cyclical	6.75%
Telecommunications	5.87%
Energy	5.36%
Utilities	3.24%
Basic Materials	1.41%
Total	99.94%

Trustee: Barclays Global Fund Advisors.
Administrator: Investors Bank & Trust Company.
Distributor: SEI Investments Distribution Co.

iShares Russell 1000 Index Fund (Symbol: IWB)

Objective

The Russell 1000 Index Fund seeks investment results that correspond generally to the price and yield performance, before fees and expenses, of the Russell 1000 Index. There is no assurance that the performance of the Russell 1000 Index can be fully matched.

Fund Details			
Options Traded:	Yes	Trading Increment:	1 Cent ($0.01)
Expense Ratio:	0.0015	Min Trade Size:	1 Share
Ticker Symbol:	IWB	Marginable:	Yes
Short Selling Allowed:	Yes, uptick exempt		

Distribution History	2000	Quick Facts	
Ordinary Income	$0.45	Net Assets:	$228,956,000
Short Term Capital Gains	$0.00	Shares Outstanding:	3,700,000
Long Term Capital Gains	$0.00	Dividend Yield:	1%
Return of Capital	$0.00	52 Week High:	$81.78
Total	$0.45	52 Week Low:	$56.98

Top Ten Holdings		Industry Groups	
1. General Electric Co.	3.97%	1. Technology	20.61%
2. Exxon Mobil Corp.	2.53%	2. Financial Services	19.63%
3. Pfizer Inc.	2.43%	3. Health Care	14.81%
4. Cisco Systems Inc	2.29%	4. Consumer Discretionary	10.51%
5. Citigroup Inc	2.15%	5. Utilities	9.01%
6. Merck & Co. Inc.	1.81%	6. Consumer Staples	6.59%
7. Intel Corp.	1.88%	7. Other	5.19%
8. American International Group Inc.	1.66%	8. Integrated Oils	4.02%
9. Microsoft Corp.	1.44%	9. Producer Durables	3.02%
10. SBC Communications Inc.	1.36%	10. Other Energy	2.38%
Total	21.32%	Total	95.77%

Trustee: Barclays Global Fund Advisors.
Administrator: Investors Bank & Trust Company.
Distributor: SEI Investments Distribution Co.

Appendix A — Fund Pages

streetTRACKS Dow Jones U.S. Large Cap Value Index Fund (Symbol: ELV)

Objective

The streetTRACKS Dow Jones U.S. Large Cap Value Index Fund seeks investment results that correspond generally to the price and yield performance, before fees and expenses, of the Dow Jones U.S. Large Cap Value Index. There is no assurance that the performance of the Dow Jones U.S. Large Cap Value Index can be fully matched.

Fund Details			
Options Traded:	No	Tracking Increment:	0.01
Expense Ratio:	0.002	Min Trade Size:	1 Share
Ticker Symbol:	ELV	Marginable:	Yes
Short Selling Allowed:	Yes, uptick exempt		
Distribution History	**2000**	**Quick Facts**	
Ordinary Income	$0.00	Net Assets:	$38,040,000
Ordinary Income	$0.57	Shares Outstanding:	300,000
Short Term Capital Gains	$0.00	Dividend Yield:	0%
Long Term Capital Gains	$0.00	52 Week High:	$138.16
Return of Capital	$0.00	52 Week Low:	$67.00
Total	$0.57	Industry Groups	
Top Ten Holdings		**Industry Groups**	
1. Exxon Mobile Corp.	5.99%	1. Financials	32.97%
2. Citigroup Inc.	5.10%	2. Healthcare	19.72%
3. Merck & Co. Inc.	4.28%	3. Consumer Non-Cyclicals	11.49%
4. American International Group Inc.	3.91%	4. Telecommunications	10.15%
5. SBC Communications	3.21%	5. Energy	6.92%
6. International Business Machines	2.97%	6. Consumer Cyclicals	6.47%
7. Bristol-Myers Squibb Co.	2.87%	7. Industrial	5.20%
8. Johnson & Johnson	2.74%	8. Technology	4.63%
9. Verizon Communications	2.68%	9. Utilities	1.27%
10. Procter & Gamble Co.	2.03%	10. Basic Materials	1.18%
Total	35.78%	Total	100%

Advisor: State Street Global Advisors
Administrator: State Street Bank & Trust Co.
Distributor: State Street Capital Markets, LLC

The Dow Industrials℠ DIAMONDS®
(Symbol: DIA)

Objective

The DIAMONDS Trust Series I is a pooled investment designed to provide investment results that generally correspond to the price and yield performance, before fees and expenses, of the Dow Jones Industrial Average (DJIA) There is no assurance that the price and yield performance of the Dow Jones Industrial Average can be fully matched.

Fund Details

Options Traded:	No	Trading Increment:	$0.01
Expense Ratio:	0.20%	Min Trade Size:	1 Share
Ticker Symbol:	DIA	Marginable:	Yes
Short Selling Allowed:	Yes, uptick exempt		

Distribution History 2000

		Quick Facts	
		Net Assets:	$166,729,500
Ordinary Income	$0.09	Shares Outstanding:	3,150,000
Short Term Capital Gains	$0.11	Dividend Yield:	0%
Long Term Capital Gains	$0.00	52 Week High:	$92.33
Return of Capital	$0.00	52 Week Low:	$47
Total	$0.20	As of Apr. 16, 2001	

Top Ten Holdings / Industry Groups

1. General Electric Co	7.72%	1. Technology	50.34%
2. Cisco Systems Inc	5.82%	2. Health Care	16.20%
3. Pfizer Inc	4.21%	3. Consumer Discretionary	12.09%
4. Intel Corp.	4.14%	4. Other	7.77%
5. Microsoft Corp	3.50%	5. Consumer Staples	3.10%
6. EMC Corp	3.21%	6. Financial Services	3.09%
7. Sun Microsystems Inc	2.75%	7. Utilities	2.67%
8. Oracle Corp	2.52%	8. Producer Durables	2.19%
9. Intl Business Machines	2.17%	9. Other Energy	2.01%
10. America Online Inc	1.84%	10. Auto & Transportation	0.34%
Total	37.89%	Total	99.80%

Trustee: Barclays Global Fund Advisors.
Administrator: Investors Bank & Trust Company.
Distributor: SEI Investments Distribution Co.

Appendix A — Fund Pages 143

iSHARES S&P 500/BARRA VALUE INDEX FUND
(SYMBOL: IVE)

Objective

The S&P 500/BARRA Value Index Fund seeks investment results the correspond generally to the price and yield performance, before fees and expenses, of the S&P 500/BARRA Value Index. There is no assurance that the performance of the S&P 500/BARRA Value Index can be fully matched.

Fund Details

Options Traded:	No	Trading Increment:	0.01
Expense Ratio:	0.0018	Min Trade Size:	1 Share
Ticker Symbol:	IVE	Marginable:	Yes
Short Selling Allowed:	Yes, uptick exempt		

Distribution History 2000 | Quick Facts

Ordinary Income	$0.45	Net Assets:	$391,950,000
Ordinary Income	$0.45	Shares Outstanding:	6,500,000
Short Term Capital Gains	$0.15	Dividend Yield:	1%
Long Term Capital Gains	$0.00	52 Week High:	$67.00
Return of Capital	$0.00	52 Week Low:	$55.00
Totals	$0.60		

Top Ten Holdings | Industry Groups

1. Exxon Mobil Corp	5.05%	1. Financial	25.48%
2. Citigroup Inc	4.29%	2. Energy	12.54%
3. SBC Communications Inc	2.70%	3. Technology	11.46%
4. Verizon Communications	2.26%	4. Communications Services	10.05%
5. Royal Dutch Petro Co	2.17%	5. Capital Goods	9.12%
6. Nortel Networks Corp	1.66%	6. Consumer Staples	8.66%
7. Philip Morris Co Inc	1.63%	7. Utilities	7.35%
8. Tyco International Ltd	1.62%	8. Consumer Cyclicals	7.03%
9. Wells Fargo & Co	1.59%	9. Basic Materials	4.59%
10. Morgan Stanley Dean Witter Discover	1.48%	10. Health Care	2.38%
Total	24.45%	Total	98.66%

Trustee: Barclays Global Fund Advisors.
Administrator: Investors Bank & Trust Company.
Distributor: SEI Investments Distribution Co.

iShares Russell 3000 Value Index Fund (Symbol: IWW)

Objective

iShares Russell 3000 Value Index Fund seeks investment results that correspond generally to the price and yield performance, before fees and expenses, of the Russell 3000 Value Index. There is no assurance that the performance of the Russell 3000 Value Index can be fully matched.

Fund Details			
Options Traded:	No	Trading Increment:	0.01
Expense Ratio:	0.0025	Min Trade Size:	1 Share
Ticker Symbol:	IWW	Marginable:	Yes
Short Selling Allowed:	Yes, uptick exempt		
Distribution History	**2000**	**Quick Facts**	
Ordinary Income	$0.45	Net Assets:	$32,391,000
Short Term Capital Gains	$0.09	Shares Outstanding:	450,000
Long Term Capital Gains	$0.00	Dividend Yield:	2%
Return of Capital	$0.00	52 Week High:	$77.25
Total	$0.64	52 Week Low:	$36.00
Top Ten Holdings		**Industry Groups**	
1. Exxon Mobil Corp	4.24%	1. Financial Services	32.41%
2. Citigroup Inc	3.61%	2. Utilities	13.81%
3. American Intl Group Inc	2.78%	3. Consumer Discretionary	9.42%
4. SBC Communications Inc	2.27%	4. Health Care	9.06%
5. Johnson & Johnson	1.94%	5. Consumer Staples	8.01%
6. Verizon Communications	1.92%	6. Integrated Oils	6.73%
7. Merck & Co.	1.52%	7. Materials & Processing	4.69%
8. Proctor & Gamble Co.	1.43%	8. Producer Durables	3.87%
9. Philip Morris Co Inc.	1.39%	9. Technology	3.55%
10. Wells Fargo & Co.	1.34%	10. Auto & Transportation	3.03%
Total	22.44%	Total	94.58%

Trustee: Barclays Global Fund Advisors.
Administrator: Investors Bank & Trust Company.
Distributor: SEI Investments Distribution Co.

Appendix A — Fund Pages

iShares Russell 1000 Value Index Fund (Symbol: IWD)

Objective

The Russell 1000 Value Index Fund seeks investment results that correspond generally to the price and yield performance, before fees and expenses, of the Russell 1000 Value Index. There is no assurance that the performance of the Russell 1000 Value Index can be fully matched.

Fund Details

Options Traded:	No	Trading Increment:	0.01
Expense Ratio:	0.002	Min Trade Size:	1 Share
Ticker Symbol:	IWD	Marginable:	Yes
Short Selling Allowed:	Yes, uptick exempt		

Distribution History — 2000

Ordinary Income	$0.45
Short Term Capital Gains	$0.00
Long Term Capital Gains	$0.00
Return of Capital	$0.00
Total	$0.44

Quick Facts

Net Assets:	$192,576,000
Shares Outstanding:	3,400,000
Dividend Yield:	1%
52 Week High:	$60.74
52 Week Low:	$51.70

Fund Details

Top Ten Holdings

1. Exxon Mobil Corp.	4.54%
2. Citigroup Inc	3.86%
3. American Intl Group Inc	2.98%
4. SBC Communications Inc	2.43%
5. Johnson & Johnson	2.08%
6. Verizon Communications	2.05%
7. Merck & Co. Inc.	1.63%
8. Proctor & Gamble Co.	1.54%
9. Philip Morris Co. Inc.	1.49%
10. Wells Fargo & Co.	1.44%
Total	24.04%

Industry Groups

1. Financial Services	31.96%
2. Utilities	15.52%
3. Health Care	9.25%
4. Consumer Discretionary	8.29%
5. Consumer Staples	7.39%
6. Integrated Oils	7.34%
7. Materials & Processing	4.86%
8. Technology	3.58%
9. Producer Durables	3.46%
10. Auto & Transportation	3.01%
Total	94.43%

Trustee: Barclays Global Fund Advisors.
Administrator: Investors Bank & Trust Company.
Distributor: SEI Investments Distribution Co.

iShares Russell 2000 Growth Index Fund (Symbol: IWO)

Objective

iShares Russell 2000 Growth Index Fund seeks investment results that correspond generally to the price and yield performance, before fees and expenses, of the Russell 2000 Growth Index. There is no assurance that the performance of the Russell 2000 Growth Index can be fully matched.

Fund Details			
Options Traded:	Yes	Trading Increment:	0.01
Expense Ratio:	0.0025	Min Trade Size:	1 Share
Ticker Symbol:	IWO	Marginable:	Yes
Short Selling Allowed:	Yes, uptick exempt		

Distribution History	2000	Quick Facts	
Ordinary Income	$0.45	Net Assets:	$157,747,500
Short Term Capital Gains	$0.04	Shares Outstanding:	2,850,000
Long Term Capital Gains	$0.00	Dividend Yield:	0%
Return of Capital	$0.00	52 Week High:	$86.50
Total	$0.06	52 Week Low:	$49.45

Top Ten Holdings		Industry Groups	
1. Caremark Rx Inc	0.78%	1. Health Care	23.98%
2. Laboratory Corp of America Holdings	0.77%	2. Technology	21.87%
3. Manugistics Group Inc	0.67%	3. Consumer Discretionary	17.82%
4. Amerisource Health Corp	0.65%	4. Financial Services	9.54%
5. Enzon Inc	0.65%	5. Producer Durables	9.10%
6. Investors Financial Services Corp.	0.64%	6. Other Energy	6.71%
7. NetIQ Corp	0.60%	7. Materials & Processing	5.31%
8. OSI Pharmaceuticals Inc.	0.58%	8. Utilities	2.09%
9. Plantronics Inc	0.58%	9. Consumer Staples	1.65%
10. Interwoven Inc	0.57%	10. Auto & Transportation	1.63%
Total	6.49%	Total	99.70%

Trustee: Barclays Global Fund Advisors.
Administrator: Investors Bank & Trust Company.
Distributor: SEI Investments Distribution Co.

IShares Russell 2000 Index Fund (Symbol: IWM)

Objective

The Russell 2000 Index Fund seeks investment results that correspond generally to the price and yield performance, before fees and expenses, of the Russell 2000 Index. There is no assurance that the performance of the Russell 2000 Index can be fully matched.

Fund Details			
Options Traded:	Yes	Trading Increment:	0.01
Expense Ratio:	0.002	Min Trade Size:	1 Share
Ticker Symbol:	IWM	Marginable:	Yes
Short Selling Allowed:	Yes, uptick exempt		
Distribution History	**2000**	**Quick Facts**	
Ordinary Income	$0.45	Net Assets:	$709,617,000
Short Term Capital Gains	$0.17	Shares Outstanding:	7,950,000
Long Term Capital Gains	$0.00	Dividend Yield:	1%
Return of Capital	$0.00	52 Week High:	$109.19
Total	$0.85	52 Week Low:	$83.15
Top Ten Holdings		**Industry Groups**	
1. Caremark Rx Inc	0.36%	1. Financial Services	22.02%
2. Laboratory Corp of America Holdings	0.35%	2. Consumer Discretionary	16.20%
3. Invitrogen Corp	0.34%	3. Health Care	14.63%
4. Health Net Inc.	0.31%	4. Technology	11.46%
5. Manugistics Group Inc	0.30%	5. Materials & Processing	9.23%
6. Amerisource Health Corp	0.30%	6. Producer Durables	8.32%
7. Enzon Inc	0.29%	7. Utilities	6.24%
8. Investors Financial Services Corp.	0.29%	8. Other Energy	4.48%
9. Astoria Financial Corp.	0.28%	9. Auto & Transportation	3.26%
10. Gallagher A J & Co.	0.28%	10. Consumer Staples	2.62%
Total	3.10%	Total	98.47%

Trustee: Barclays Global Fund Advisors.
Administrator: Investors Bank & Trust Company.
Distributor: SEI Investments Distribution Co.

IShares Russell 2000 Value Index Fund (Symbol: IWN)

Objective

iShares Russell 2000 Value Index Fund seeks investment results that correspond generally to the price and yield performance, before fees and expenses, of the Russell 2000 Value Index. There is no assurance that the performance of the Russell 2000 Value Index can be fully matched.

Fund Details			
Options Traded:	Yes	Trading Increment:	0.01
Expense Ratio:	0.0025	Min Trade Size:	1 Share
Ticker Symbol:	IWN	Marginable:	Yes
Short Selling Allowed:	Yes, uptick exempt		

Distribution History	2000	Quick Facts	
Ordinary Income	$0.45	Net Assets:	$272,717,500
Short Term Capital Gains	$0.03	Shares Outstanding:	2,350,000
Long Term Capital Gains	$0.00	Dividend Yield:	0%
Return of Capital	$0.00	52 Week High:	$123.28
Total	$0.89	52 Week Low:	$99.50

Top Ten Holdings		Industry Groups	
1. Health Net Inc.	0.57%	1. Financial Services	32.54%
2. Astoria Financial Corp	0.52%	2. Consumer Discretionary	14.88%
3. Humana Inc	0.50%	3. Materials & Processing	12.56%
4. Centex Corp	0.47%	4. Utilities	9.70%
5. Bank United Corp Class A	0.46%	5. Producer Durables	7.67%
6. Venator Group Inc	0.45%	6. Health Care	6.78%
7. Bergen Brunswig Corp Class A	0.45%	7. Auto & Transportation	4.66%
8. Pactiv Corp	0.44%	8. Consumer Staples	3.44%
9. Precision Castparts Corp	0.44%	9. Technology	2.75%
10. Mdu Resources Group Inc	0.44%	10. Other Energy	2.60%
Total	4.74%	Total	97.58%

Trustee: Barclays Global Fund Advisors.
Administrator: Investors Bank & Trust Company.
Distributor: SEI Investments Distribution Co.

FINANCIAL

THE FINANCIAL SELECT SECTOR SPDR FUND (SYMBOL: XLF)

Objective

The Financial Select Sector SPDR Fund is one of nine select sector SPDR funds. Each fund is designed to accumulate and hold a portfolio of common stocks intended to provide investment results that generally correspond to the price performance and yield, before fees and expenses, of a specific select sector index. There is no assurance that the price and yield performance of a specific select sector index can be fully matched. Collectively the nine select sector indexes represent all of the companies in the S&P 500 Index. As of 12/29/00 the Financial Select Sector SPDR Fund held 73 stocks whose weight within the S&P 500 Index was approximately 17.31 percent.

Fund Details			
Options Traded:	Yes	Trading Increment:	$0.01
Expense Ratio:	0.28%	Min Trade Size:	1 Share
Ticker Symbol:	XLF	Marginable:	Yes
Short Selling Allowed:	Yes, uptick exempt		
Distribution History	**2000**	**Quick Facts**	
Ordinary Income	$0.36	Net Assets:	$720,217,500
Short Term Capital Gains	$0.00	Shares Outstanding:	27,250,000
Long Term Capital Gains	$0.00	Dividend Yield:	2%
Return of Capital	$0.00	52 Week High:	$30.66
Totals	$0.36	52 Week Low:	$22.12
Top Ten Holdings		**Industry Groups**	
1. Citigroup Inc.	11.87%	1. Financial (Diversified)	29.45%
2. American International Group Inc.	11.64%	2. Banks (Major Regional)	23.30%
3. Wells Fargo and Co.	10.13%	3. Insurance (Multi-Line)	12.52%
4. Morgan Stanley Dean Witter & Co.	5.23%	4. Banks (Money Center)	7.91%
5. Federal Nat'l Mtg. Assn.	4.92%	5. Invest. Banking/Brokerage	5.69%
6. Bank of America Corp.	4.87%	6. Insurance (Life/Health)	4.85%
7. American Express Co.	4.22%	7. Consumer Finance	4.53%
8. J. P. Morgan Chase & Co.	3.44%	8. Insurance (Property - Casualty)	4.49%
9. Merrill Lynch and Co. Inc.	3.42%	9. Savings & Loan Companies	2.22%
10. Federal Home Ln Mtg Corp	3.18%	10. Insurance Brokers	2.00%
Total	62.91%	Total	96.96%

Advisor: State Street Bank
Sponsor: PDR Services LLC
Distributor: ALPS Mutual Funds Inc.

iShares Dow Jones U.S. Financial Sector Index Fund (Symbol: IYF)

Objective

The Dow Jones U.S. Financial Sector Index Fund seeks investment results that correspond to the price and yield performance, before fees and expenses, of the Dow Jones U.S. Financial Sector Index. There is no assurance that the performance of the Dow Jones U.S. Financial Sector Index can be fully matched.

Fund Details			
Options Traded:	No	Trading Increment:	$0.01
Expense Ratio:	0.60%	Min Trade Size:	1 Share
Ticker Symbol:	IYF	Marginable:	Yes
Short Selling Allowed:	Yes, uptick exempt		

Distribution History	2000	Quick Facts	
		Net Assets:	$59,280,000
Ordinary Income	$1.44	Shares Outstanding:	750,000
Short Term Capital Gains	$0.39	Dividend Yield:	5%
Long Term Capital Gains	$0.00	52 Week High:	$90.00
Return of Capital	$0.00	52 Week Low:	$70.47
Totals	$1.83	As of Apr. 16, 2001	

Top Ten Holdings		Industry Groups	
1. Citigroup Inc.	10.83%	1. Specialty Finance	44.26%
2. American International Group	8.51%	2. Banks	32.94%
3. Morgan Stanley Dean Witter Discover	4.61%	3. Insurance	18.03%
4. Bank of America Corp	3.83%	4. Real Estate	4.64%
5. Wells Fargo & Co	3.37%		
6. Federal Natl Mtg Assn	3.25%		
7. American Express Co	3.19%		
8. Chase Manhattan Bank Corp	2.68%		
9. Merrill Lynch & Co	2.12%		
10. Bank One Corp	1.99%		
Total	44.37%	Total	99.87%

Trustee: Barclays Global Fund Advisors.
Administrator: Investors Bank & Trust Company.
Distributor: SEI Investments Distribution Co.

Appendix A — Fund Pages   151

iShares Dow Jones U.S. Financial Services Sector Index Fund (Symbol: IYG)

Objective

The Dow Jones U.S. Financial Services Sector Index Fund seeks investment results that correspond to the price and yield performance, before fees and expenses, of the Dow Jones U.S. Financial Services Sector Index. There is no assurance that the performance of the Dow Jones U.S. Financial Services Sector Index can be fully matched.

Fund Details			
Options Traded:	No	Trading Increment:	$0.01
Expense Ratio:	0.60%	Min Trade Size:	1 Share
Ticker Symbol:	IYG	Marginable:	Yes
Short Selling Allowed:	Yes, uptick exempt		

Distribution History	2000	Quick Facts	
		Net Assets:	$27,195,000
Ordinary Income	$0.93	Shares Outstanding:	300,000
Short Term Capital Gains	$0.13	Dividend Yield:	2%
Long Term Capital Gains	$0.00	52 Week High:	$104.22
Return of Capital	$0.00	52 Week Low:	$40.50
Totals	$1.06	As of Apr. 16, 2001	

Top Ten Holdings		Industry Groups	
1. Citigroup Inc.	14.01%	1. Specialty Finance	57.29%
2. Morgan Stanley Dean Witter Discover	5.96%	2. Banks	42.62%
3. Bank of America Corp	4.96%		
4. Wells Fargo & Co	4.36%		
5. Federal Natl Mtg Assn	4.20%		
6. American Express Co	4.13%		
7. Chase Manhattan Corp	3.47%		
8. Merrill Lynch & Co Inc	2.74%		
9. Bank One Corp	2.57%		
10. Bank of New York Inc	2.38%		
Total	48.79%	Total	99.91%

Trustee: Barclays Global Fund Advisors.
Administrator: Investors Bank & Trust Company.
Distributor: SEI Investments Distribution Co.

HEALTH CARE

iShares Nasdaq Biotechnology Index Fund
(Symbol: IBB)

Objective

The iShares Nasdaq Biotechnology Index Fund seeks investment results that correspond generally to the price and yield performance, before fees and expenses, of the Nasdaq Biotechnology Index. There is no assurance that the performance of the Nasdaq Biotechnology Index can be fully matched.

Fund Details			
Options Traded:	No	Trading Increment:	$0.01
Expense Ratio:	0.50%	Min Trade Size:	1 Share
Ticker Symbol:	IBB	Marginable:	Yes
Short Selling Allowed:	Yes, uptick exempt		

Distribution History	2000	Quick Facts	
None at present		Net Assets:	$
		Shares Outstanding:	0
		Dividend Yield:	0%
		52 Week High:	$106.20
		52 Week Low:	$61.00

Top Ten Holdings		Industry Groups	
1. Amgen Inc	27.70%	1. Biotechnology	100.00%
2. Immunex Corporation	8.80%		
3. Millenium Pharmaceuticals, Inc	5.50%		
4. Medimmune, Inc	4.20%		
5. Biogen, Inc	3.70%		
6. IDEC Pharmaceuticals Corp	3.60%		
7. Chiron Corporation	3.60%		
8. Genzyme General	3.30%		
9. Human Genome Sciences, Inc	3.20%		
10. Abgenix, Inc	2.00%		
Total	65.60%	Total	100.00%

Trustee: Barclays Global Fund Advisors.
Administrator: Investors Bank & Trust Company.
Distributor: SEI Investments Distribution Co.

iShares Dow Jones U.S. Healthcare Sector Index Fund (Symbol: IYH)

Objective

The Dow Jones U.S. Healthcare Sector Index Fund seeks investment results that correspond to the price and yield performance, before fees and expenses, of the Dow Jones U.S. Healthcare Sector Index. There is no assurance that the performance of the Dow Jones U.S. Healthcare Sector Index can be fully matched.

Fund Details			
Options Traded:	No	Trading Increment:	$0.01
Expense Ratio:	0.60%	Min Trade Size:	1 Share
Ticker Symbol:	IYH	Marginable:	Yes
Short Selling Allowed:	Yes, uptick exempt		
Distribution History	**2000**	**Quick Facts**	
		Net Assets:	$83,300,000
Ordinary Income	$0.07	Shares Outstanding:	1,400,000
Short Term Capital Gains	$0.09	Dividend Yield:	0%
Long Term Capital Gains	$0.00	52 Week High:	$73.00
Return of Capital	$0.00	52 Week Low:	$52.00
Totals	$0.16	As of Apr. 16, 2001	
Top Ten Holdings		**Industry Groups**	
1. Pfizer Inc	16.84%	1. Pharmaceutical and Biotechnology	79.20%
2. Merck & Co Inc	10.20%	2. Medical Products	14.46%
3. Johnson & Johnson	7.36%	3. Healthcare Providers	6.31%
4. Bristol Myers Squibb Co	6.69%		
5. Lilly (Eli) & Co	4.64%		
6. Pharmacia Corp	4.55%		
7. American Home Prods Corp	4.39%		
8. Amgen Inc	4.27%		
9. Abbott Labs	4.10%		
10. Schering-Plough Corp	4.06%		
Total	67.10%	Total	99.97%

Trustee: Barclays Global Fund Advisors.
Administrator: Investors Bank & Trust Company.
Distributor: SEI Investments Distribution Co.

BASIC INDUSTRY

iShares Dow Jones U.S. Basic Materials Sector Index Fund (Symbol: IYM)

Objective

The Dow Jones U.S. Basic Materials Sector Index Fund seeks investment results that correspond to the price and yield performance, before fees and expenses, of the Dow Jones U.S. Basic Materials Sector Index. There is no assurance that the performance of the Dow Jones U.S. Basic Materials Sector Index can be fully matched.

Fund Details			
Options Traded:	No	Trading Increment:	$0.01
Expense Ratio:	0.60%	Min Trade Size:	1 Share
Ticker Symbol:	IYM	Marginable:	Yes
Short Selling Allowed:	Yes, uptick exempt		
Distribution History	**2000**	**Quick Facts**	
		Net Assets:	$9,432,500
Ordinary Income	$0.39	Shares Outstanding:	250,000
Short Term Capital Gains	$0.07	Dividend Yield:	2%
Long Term Capital Gains	$0.00	52 Week High:	$40.65
Return of Capital	$0.00	52 Week Low:	$29.56
Total	$0.46	As of Apr. 16, 2001	
Top Ten Holdings		**Industry Groups**	
1. Du Pont E I De Nemours & Co	21.52%	1. Chemicals	59.15%
2. Alcoa Inc	10.91%	2. Forest Products & Paper	21.30%
3. Dow Chem Co	8.42%	3. Mining & Metals	19.42%
4. Intl Paper Co	6.91%		
5. Weyerhaeuser Co	4.30%		
6. Air Products & Chemicals Inc	4.11%		
7. Praxair Inc	2.94%		
8. Avery Dennison Corp	2.58%		
9. Union Carbide Corp	2.54%		
10. Rohm & Haas Co	2.19%		
Total	66.42%	Total	99.87%

Trustee: Barclays Global Fund Advisors.
Administrator: Investors Bank & Trust Company.
Distributor: SEI Investments Distribution Co.

BASIC INDUSTRIES SELECT SECTOR SPDR

Select Sector SPDR Fund
(Symbol: XLB)

Objective

The Basic Industries Select Sector SPDR Fund is one of nine select sector SPDR funds. Each fund is designed to accumulate and hold a portfolio of common stocks intended to provide investment results that generally correspond to the price performance and yield, before fees and expenses, of a specific select sector index. There is no assurance that the price and yield performance of a specific select sector index can be fully matched. Collectively the nine select sector indexes represent all of the companies in the S&P 500 Index. As of 12/29/00 the Basic Industries Select Sector SPDR Fund held 44 stocks whose weight within the S&P 500 Index was approximately 2.38 percent.

Fund Details			
Options Traded:	Yes	Trading Increment:	$0.01
Expense Ratio:	0.28%	Min Trade Size:	1 Share
Ticker Symbol:	XLB	Marginable:	Yes
Short Selling Allowed:	Yes, uptick exempt		
Distribution History	**2000**	**Quick Facts**	
Ordinary Income	$0.40	Net Assets:	$81,282,000
Short Term Capital Gains	$0.51	Shares Outstanding:	3,800,000
Long Term Capital Gains	$0.00	Dividend Yield:	2%
Return of Capital	$0.00	52 Week High:	$24.37
Totals	$0.90	52 Week Low:	$17.00
Top Ten Holdings		**Industry Groups**	
1. Du Pont E I de Nemours	17.96%	1. Chemicals	39.45%
2. Alcoa Inc.	10.33%	2. Paper & Forest Products	18.87%
3. Dow Chemical Co.	8.86%	3. Aluminum	14.27%
4. International Paper Co.	7.00%	4. Gold/Precious Metals Mining	4.84%
5. Weyerhaeuser Co.	3.97%	5. Chemicals - Diversified	4.50%
6. Alcan Alum Ltd.	3.94%	6. Chemicals Speciality	4.43%
7. Air Products and Chemicals Inc.	3.35%	7. Metals Mining	3.10%
8. Rohm and Haas Co.	2.85%	8. Manufacturing (Specialized)	3.06%
9. PPG Industries Inc.	2.79%	9. Steel	2.37%
10. Union Carbide Corp	2.59%	10. Containers/Packaging (Paper)	2.27%
Total	63.64%	Total	97.17%

Advisor: State Street Bank
Sponsor: PDR Services LLC
Distributor: ALPS Mutual Funds Inc.

The Industrial Select Sector SPDR Fund
(Symbol: XLI)

Objective

The Industrial Select Sector SPDR Fund is one of 9 select sector SPDR funds. Each fund is designed to accumulate and hold a portfolio of common stocks intended to provide investment results that generally correspond to the price performance and yield, before fees and expenses, of a specific select sector index. There is no assurance that the price and yield performance of a specific select sector index can be fully matched. Collectively the nine select sector index represent all of the companies in the S&P 500 Index. As of 12/29/00 the Industrial Select Sector SPDR Fund held 44 stocks whose weight within the S&P 500 Index was approximately 9.15 percent.

Fund Details

Options Traded:	Yes	Trading Increment:	$0.01
Expense Ratio:	0.28%	Min Trade Size:	1 Share
Ticker Symbol:	XLF	Marginable:	Yes
Short Selling Allowed:	Yes, uptick exempt		

Distribution History 2000

Ordinary Income	$0.36
Short Term Capital Gains	$0.00
Long Term Capital Gains	$0.00
Return of Capital	$0.00
Totals	$0.36

Quick Facts

Net Assets:	$720,217,500
Shares Outstanding:	27,250,000
Dividend Yield:	2%
52 Week High:	$30.66
52 Week Low:	$22.12

Top Ten Holdings

1. Citigroup Inc.	11.87%
2. American International Group Inc.	11.64%
3. Wells Fargo and Co.	10.13%
4. Morgan Stanley Dean Witter & Co.	5.23%
5. Federal Nat'l Mtg. Assn.	4.92%
6. Bank of America Corp.	4.87%
7. American Express Co.	4.22%
8. J. P. Morgan Chase & Co.	3.44%
9. Merrill Lynch and Co. Inc.	3.42%
10. Federal Home Ln Mtg Corp	3.18%
Total	62.91%

Industry Groups

1. Financial (Diversified)	29.45%
2. Banks (Major Regional)	23.30%
3. Insurance (Multi-Line)	12.52%
4. Banks (Money Center)	7.91%
5. Invest. Banking/Brokerage	5.69%
6. Insurance (Life/Health)	4.85%
7. Consumer Finance	4.53%
8. Insurance (Property - Casualty)	4.49%
9. Savings & Loan Companies	2.22%
10. Insurance Brokers	2.00%
Total	96.96%

Advisor: State Street Bank
Sponsor: PDR Services LLC
Distributor: ALPS Mutual Funds Inc.

Appendix A — Fund Pages

iShares Dow Jones U.S. Industrial Sector Index Fund (Symbol: IYJ)

Objective

The Dow Jones U.S. Industrial Sector Index Fund seeks investment results that correspond to the price and yield performance, before fees and expenses, of the Dow Jones U.S. Industrial Sector Index. There is no assurance that the performance of the Dow Jones U.S. Industrial Sector Index can be fully matched.

Fund Details			
Options Traded:	No	Trading Increment:	$0.01
Expense Ratio:	0.60%	Min Trade Size:	1 Share
Ticker Symbol:	IYJ	Marginable:	Yes
Short Selling Allowed:	Yes, uptick exempt		

Distribution History	2000	Quick Facts	
		Net Assets:	$22,005,000
Ordinary Income	$0.15	Shares Outstanding:	450,000
Short Term Capital Gains	$0.13	Dividend Yield:	1%
Long Term Capital Gains	$0.00	52 Week High:	$64.72
Return of Capital	$0.00	52 Week Low:	$44.85
Total	$0.28	As of Apr. 16, 2001	

Top Ten Holdings		Industry Groups	
1. General Elec Co	22.54%	1. Industrial, Diversified	40.15%
2. Tyco International	5.28%	2. Industrial Equipment	22.58%
3. Corning Inc	5.07%	3. Industrial Services	21.36%
4. JDS Uniphase Corp	4.05%	4. Aerospace	8.53%
5. Boeing Co	3.58%	5. Industrial Transportation	3.96%
6. Minnesota Mng & Mfg Co	3.36%	6. Construction and Materials	2.08%
7. Automatic Data Processing	3.00%	7. Containers & Packaging	0.94%
8. Emerson Electric Co	2.31%	8. Transportation Equipment	0.31%
9. United Technologies Corp	2.29%		
10. Honeywell International Inc	1.70%		
Total	53.19%	Total	99.91%

Trustee: Barclays Global Fund Advisors.
Administrator: Investors Bank & Trust Company.
Distributor: SEI Investments Distribution Co.

UTILITIES/ENERGY

THE ENERGY SELECT SECTOR SPDR FUND
(SYMBOL: XLE)

Objective

The Energy Select Sector SPDR Fund is one of nine select sector SPDR funds. Each fund is designed to accumulate and hold a portfolio of common stocks intended to provide investment results that generally correspond to the price performance and yield, before fees and expenses, of a specific select sector index. There is no assurance that the price and yield performance of a specific select sector index can be fully matched. Collectively the nine select sector indexes represent all of the companies in the S&P 500 Index. As of 12/29/00 the Energy Select Sector SPDR Fund held 31 stocks whose weight within the S&P 500 Index was approximately 7.44 percent.

Fund Details			
Options Traded:	Yes	Trading Increment:	$0.01
Expense Ratio:	0.28%	Min Trade Size:	1 Share
Ticker Symbol:	XLE	Marginable:	Yes
Short Selling Allowed:	Yes, uptick exempt		
Distribution History	**2000**	**Quick Facts**	
Ordinary Income	$0.48	Net Assets:	$227,574,000
Short Term Capital Gains	$0.00	Shares Outstanding:	7,050,000
Long Term Capital Gains	$0.00	Dividend Yield:	1%
Return of Capital	$0.00	52 Week High:	$34.75
Totals	$0.48	52 Week Low:	$27.51
Top Ten Holdings		**Industry Groups**	
1. Exxon Mobil Corp.	11.87%	1. Oil (Int'l Integrated)	48.86%
2. Royal Dutch Petroleum	11.64%	2. Natural Gas - Distr - Pipe Line	16.19%
3. Enron Corp.	10.13%	3. Oil & Gas (Drilling & Equip)	12.40%
4. Texaco Inc.	5.23%	4. Oil & Gas (Exploration/Prod)	11.68%
5. Schlumberger Ltd.	4.92%	5. Oil (Domestic Integrated)	6.81%
6. Chevron Corp.	4.87%	6. Oil & Gas (Refining & Mktg)	3.17%
7. Coastal Corp	4.22%	7. Engineering & Construction	0.67%
8. Conoco Inc.	3.44%	8. Other	0.22%
9. Anadarko Pete Corp.	3.42%		
10. El Paso Energy Corp. Del	3.18%		
Total	62.91%	Total	100.00%

Advisor: State Street Bank
Sponsor: PDR Services LLC
Distributor: ALPS Mutual Funds Inc.

iShares Dow Jones U.S. Energy Sector Index Fund (Symbol: IYE)

Objective

The Dow Jones U.S. Energy Sector Index Fund seeks investment results that correspond to the price and yield performance, before fees and expenses, of the Dow Jones U.S. Energy Sector Index. There is no assurance that the performance of the Dow Jones U.S. Energy Sector Index can be fully matched.

Fund Details

Options Traded:	No	Trading Increment:	$0.01
Expense Ratio:	0.60%	Min Trade Size:	1 Share
Ticker Symbol:	IYE	Marginable:	Yes
Short Selling Allowed:	Yes, uptick exempt		

Distribution History 2000

Ordinary Income	$0.27
Short Term Capital Gains	$0.00
Long Term Capital Gains	$0.00
Return of Capital	$0.00
Totals	$0.27

Quick Facts

Net Assets:	$41,688,000
Shares Outstanding:	800,000
Dividend Yield:	1%
52 Week High:	$57.59
52 Week Low:	$45.37
As of Apr. 16, 2001	

Top Ten Holdings

1. Exxon Mobil Corp	22.99%
2. Chevron Corp	10.32%
3. Schlumberger Ltd	6.53%
4. Texaco Inc	5.39%
5. Halliburton Co	3.31%
6. Phillips Petroleum Co	2.91%
7. Williams Co Inc	2.65%
8. Anadarko Petroleum Corp	2.42%
9. Murphy Oil Corp	2.39%
10. Conoco Inc Class B	2.28%
Total	61.19%

Industry Groups

1. Oil & Gas	99.98%
Total	99.98%

Trustee: Barclays Global Fund Advisors.
Administrator: Investors Bank & Trust Company.
Distributor: SEI Investments Distribution Co.

The Utilities Select Sector SPDR Fund (Symbol: XLU)

Objective

The Utilities Select Sector SPDR Fund is one of nine select sector SPDR funds. Each fund is designed to accumulate and hold a portfolio of common stocks intended to provide investment results that generally correspond to the price performance and yield, before fees and expenses, of a specific select sector index. There is no assurance that the price and yield performance of a specific select sector index can be fully matched. Collectively the nine select sector indexes represent all of the companies in the S&P 500 Index. As of 12/29/00 the Utilities Select Sector SPDR Fund held 39 stocks whose weight within the S&P 500 Index was approximately 6.26 percent.

Fund Details

Options Traded:	Yes	Trading Increment:	$0.01
Expense Ratio:	0.28%	Min Trade Size:	1 Share
Ticker Symbol:	XLU	Marginable:	Yes
Short Selling Allowed:	Yes, uptick exempt		

Distribution History	2000	Quick Facts	
Ordinary Income	$0.91	Net Assets:	$81,447,000
Short Term Capital Gains	$0.25	Shares Outstanding:	2,550,000
Long Term Capital Gains	$0.00	Dividend Yield:	3%
Return of Capital	$0.00	52 Week High:	$34.55
Totals	$1.16	52 Week Low:	$26.00

Top Ten Holdings		Industry Groups	
1. SBC Communications Inc.	11.87%	1. Electric Companies	46.41%
2. Verizon Communications	11.64%	2. Telephone	41.37%
3. Duke Energy Corp.	10.13%	3. Natural Gas - Distr - Pipe Line	6.88%
4. Exelon Corp	5.23%	4. Power Producers (Independ.)	5.12%
5. AES Corp.	4.92%	5. Other	0.22%
6. Southern Co.	4.87%		
7. Bellsouth Corp.	4.22%		
8. American Electric Power Co Inc	3.44%		
9. Alltel Corp	3.42%		
10. Dominion Res Inc Va New	3.18%		
Total	62.91%	Total	100.00%

Advisor: State Street Bank
Sponsor: PDR Services LLC
Distributor: ALPS Mutual Funds Inc.

Appendix A — Fund Pages

iSHARES DOW JONES U.S. UTILITIES SECTOR INDEX FUND (SYMBOL: IDU)

Objective

The Dow Jones U.S. Utilities Sector Index Fund seeks investment results that correspond to the price and yield performance, before fees and expenses, of the Dow Jones U.S. Utilities Sector Index. There is no assurance that the performance of the Dow Jones U.S. Utilities Sector Index can be fully matched.

Fund Details

Options Traded:	No	Trading Increment:	$0.01
Expense Ratio:	0.60%	Min Trade Size:	1 Share
Ticker Symbol:	IDU	Marginable:	Yes
Short Selling Allowed:	Yes, uptick exempt		

Distribution History	2000	Quick Facts	
		Net Assets:	$28,997,500
Ordinary Income	$1.12	Shares Outstanding:	350,000
Short Term Capital Gains	$0.14	Dividend Yield:	3%
Long Term Capital Gains	$0.00	52 Week High:	$89.75
Return of Capital	$0.00	52 Week Low:	$65.78
Total	$1.26	As of Apr. 16, 2001	

Top Ten Holdings		Industry Groups	
1. Enron Corp	14.18%	1. Electric Utilities	94.19%
2. Duke Power Corp	6.91%	2. Gas Utilities	5.00%
3. AES Corp	5.60%	3. Water Utilities	0.75%
4. Southern Co	4.61%		
5. Calpine Corp	3.15%		
6. Dominion Res Inc	3.02%		
7. American Electric Power Inc	2.76%		
8. Reliant Energy Inc	2.55%		
9. FPL Group Inc	2.32%		
10. TXU Corp	2.29%		
Total	47.39%	Total	99.94%

Trustee: Barclays Global Fund Advisors.
Administrator: Investors Bank & Trust Company.
Distributor: SEI Investments Distribution Co.

CHEMICALS

iShares Dow Jones U.S. Chemical Sector Index Fund (Symbol: IYD)

Objective

The Dow Jones U.S. Chemical Sector Index Fund seeks investment results that correspond to the price and yield performance, before fees and expenses, of the Dow Jones U.S. Chemical Sector Index. There is no assurance that the performance of the Dow Jones U.S. Chemical Sector Index can be fully matched.

Fund Details			
Options Traded:	No	Trading Increment:	$0.01
Expense Ratio:	0.60%	Min Trade Size:	1 Share
Ticker Symbol:	IYD	Marginable:	Yes
Short Selling Allowed:	Yes, uptick exempt		

Distribution History	2000	Quick Facts	
		Net Assets:	$16,780,000
Ordinary Income	$0.49	Shares Outstanding:	400,000
Short Term Capital Gains	$0.25	Dividend Yield:	2%
Long Term Capital Gains	$0.00	52 Week High:	$44.85
Return of Capital	$0.00	52 Week Low:	$18.50
Total	$0.74	As of Apr. 16, 2001	

Top Ten Holdings		Industry Groups	
1. Du Pont (E I) De Nemours & Co	22.46%	1. Chemicals	99.74%
2. Dow Chemical Co	13.77%		
3. Air Prods & Chem Co	6.70%		
4. Praxair Inc	4.95%		
5. Avery Dennison Corp	4.55%		
6. Union Carbide Corp	4.53%		
7. Rohm & Haas Co	3.81%		
8. Ecolab Inc	3.74%		
9. Eastman Chem Co	3.00%		
10. Sigma Aldrich Corp	2.66%		
Total	70.18%	Total	99.74%

Trustee: Barclays Global Fund Advisors.
Administrator: Investors Bank & Trust Company.
Distributor: SEI Investments Distribution Co.

# CONSUMER

The Consumer Services Select Sector SPDR Fund
(Symbol: XLV)

Objective

The Consumer Services Select Sector SPDR Fund is one of nine select sector SPDR funds. Each fund is designed to accumulate and hold a portfolio of common stocks intended to provide investment results that generally correspond to the price performance and yield, before fees and expenses, of a specific select sector index. There is no assurance that the price and yield performance of a specific select sector index can be fully matched. Collectively the nine select sector indexes represent all of the companies in the S&P 500 Index. As of 12/29/00 the Consumer Services Select Sector SPDR Fund held 41 stocks whose weight within the S&P 500 Index was approximately 5.04 percent.

Fund Details

Options Traded:	Yes	Trading Increment:	$0.01
Expense Ratio:	0.28%	Min Trade Size:	1 Share
Ticker Symbol:	XLV	Marginable:	Yes
Short Selling Allowed:	Yes, uptick exempt		

Distribution History	2000	Quick Facts	
Ordinary Income	$0.05	Net Assets:	$103,702,000
Short Term Capital Gains	$0.01	Shares Outstanding:	3,800,000
Long Term Capital Gains	$0.00	Dividend Yield:	1%
Return of Capital	$0.00	52 Week High:	$30.66
Totals	$0.06	52 Week Low:	$24.55

Top Ten Holdings		Industry Groups	
1. Viacom Inc.	11.87%	1. Entertainment	33.64%
2. Time Warner Inc.	11.64%	2. Broadcasting (TV, Radio & Cable)	10.10%
3. Walt Disney Co.	10.13%	3. Health Care (Managed Care)	9.92%
4. Comcast Corp.	5.23%	4. Restaurants	8.59%
5. McDonalds Corp.	4.92%	5. Publishing - Newspapers	8.05%
6. Clear Channel Communications	4.87%	6. Lodging - Hotels	6.89%
7. HCA - The Healthcare Company	4.22%	7. Health Care (Hospital Mgmt)	6.74%
8. UnitedHealth Group Inc.	3.44%	8. Services (Advertising/Mktg)	4.86%
9. Cigna Corp.	3.42%	9. Services (Commercial & Consum)	3.50%
10. Carnival Corp.	3.18%	10. Publishing	2.38%
Total	62.91%	Total	94.67%

Advisor: State Street Bank
Sponsor: PDR Services LLC
Distributor: ALPS Mutual Funds Inc.

THE CONSUMER STAPLES SELECT SECTOR SPDR FUND (SYMBOL: XLP)

Objective

The Consumer Staples Select Sector SPDR Fund is one of nine select sector SPDR funds. Each fund is designed to accumulate and hold a portfolio of common stocks intended to provide investment results that generally correspond to the price performance and yield, before fees and expenses, of a specific select sector index. There is no assurance that the price and yield performance of a specific select sector index can be fully matched. Collectively the nine select sector indexes represent all of the companies in the S&P 500 Index. As of 12/29/00 the Consumer Staples Select Sector SPDR Fund held 68 stocks whose weight within the S&P 500 Index was approximately 21.42 percent.

Fund Details			
Options Traded:	Yes	Trading Increment:	$0.01
Expense Ratio:	0.28%	Min Trade Size:	1 Share
Ticker Symbol:	XLP	Marginable:	Yes
Short Selling Allowed:	Yes, uptick exempt		
Distribution History	**2000**	**Quick Facts**	
Ordinary Income	$0.30	Net Assets:	$207,825,000
Short Term Capital Gains	$0.00	Shares Outstanding:	8,500,000
Long Term Capital Gains	$0.00	Dividend Yield:	1%
Return of Capital	$0.00	52 Week High:	$28.87
Totals	$0.30	52 Week Low:	$21.55
Top Ten Holdings		**Industry Groups**	
1. Pfizer Inc.	11.87%	1. Health Care (Drugs/Pharms)	31.69%
2. Merck and Company Inc.	11.64%	2. Health Care Diversified	17.85%
3. Coca Cola Co.	10.13%	3. Beverages (Non-Alcoholic)	9.17%
4. Johnson and Johnson	5.23%	4. Household Prods (Non-Durable)	7.37%
5. Bristol Myers Squibb Co.	4.92%	5. Foods	6.56%
6. Lilly Eli and Company	4.87%	6. Health Care (Medical Prods/Sups)	6.21%
7. Procter and Gamble Co.	4.22%	7. Tobacco	4.07%
8. Philip Morris Cos Inc	3.44%	8. Biotechnology	3.72%
9. American Home Products	3.42%	9. Retail Stores - Food Chains	2.68%
10. Schering Plough Corp	3.18%	10. Retail Stores - Drug Stores	2.65%
Total	62.91%	Total	91.98%

Advisor: State Street Bank
Sponsor: PDR Services LLC
Distributor: ALPS Mutual Funds Inc.

TECHNOLOGY

iShares Dow Jones U.S. Technology Sector Index Fund (Symbol: IYW)

Objective

The Dow Jones U.S. Technology Sector Index Fund seeks investment results that correspond to the price and yield performance, before fees and expenses, of the Dow Jones U.S. Technology Sector Index. There is no assurance that the performance of the Dow Jones U.S. Technology Sector Index can be fully matched.

Fund Details

Options Traded:	No	Trading Increment:	$0.01
Expense Ratio:	0.60%	Min Trade Size:	1 Share
Ticker Symbol:	IYW	Marginable:	Yes
Short Selling Allowed:	Yes, uptick exempt		

Distribution History — 2000

Ordinary Income	$0.00
Short Term Capital Gains	$0.00
Long Term Capital Gains	$0.00
Return of Capital	$0.00
Total	$0.00

Quick Facts

Net Assets:	$94,843,000
Shares Outstanding:	1,700,000
Dividend Yield:	0%
52 Week High:	$139.00
52 Week Low:	$46.20
As of Apr. 16, 2001	

Top Ten Holdings

1. Cisco Systems Inc	10.33%
2. Intel Corp	7.03%
3. Microsoft Corp	6.72%
4. EMC Corp	5.75%
5. Intl. Business Machines	5.27%
6. Sun Microsystems Inc	4.93%
7. Oracle Corp	4.56%
8. Lucent Technologies Inc	2.72%
9. Texas Instruments Inc	2.17%
10. Hewlett Packard Co	2.15%
Total	51.61%

Industry Groups

1. Hardware and Equipment	74.73%
2. Software	25.28%
Total	100.01%

Trustee: Barclays Global Fund Advisors.
Administrator: Investors Bank & Trust Company.
Distributor: SEI Investments Distribution Co.

The Technology Select Sector SPDR Fund (Symbol: XLK)

Objective

The Technology Select Sector SPDR Fund is one of nine select sector SPDR funds. Each fund is designed to accumulate and hold a portfolio of common stocks intended to provide investment results that generally correspond to the price performance and yield, before fees and expenses, of a specific select sector index. There is no assurance that the price and yield performance of a specific sSelect sector index can be fully matched. Collectively the nine select sector indexes represent all of the companies in the S&P 500 Index. As of 12/29/00 the Technology Select Sector SPDR Fund held 94 stocks whose weight within the S&P 500 Index was approximately 24.08 percent.

Fund Details

Options Traded:	Yes	Trading Increment:	$0.01
Expense Ratio:	0.28%	Min Trade Size:	1 Share
Ticker Symbol:	XLK	Marginable:	Yes
Short Selling Allowed:	Yes, uptick exempt		

Distribution History	2000	Quick Facts	
Ordinary Income	$0.00	Net Assets:	$1,070,965,000
Short Term Capital Gains	$0.00	Shares Outstanding:	41,350,000
Long Term Capital Gains	$0.00	Dividend Yield:	0%
Return of Capital	$0.00	52 Week High:	$57.75
Totals	$0.00	52 Week Low:	$21.85

Top Ten Holdings		Industry Groups	
1. Cisco Systems Inc.	11.87%	1. Computers Software/Services	22.36%
2. Microsoft Corp.	11.64%	2. Electronics - Semiconductors	15.32%
3. Intel Corp.	10.13%	3. Computers (Hardware)	14.24%
4. Oracle Corporation	5.23%	4. Communications Equipment	12.47%
5. Int'l Business Machines	4.92%	5. Computers (Networking)	10.68%
6. EMC Corporation	4.87%	6. Computers (Peripherals)	5.35%
7. Nortel Networks Corp.	4.22%	7. Telephone Long Distance	4.84%
8. Sun Microsystems	3.44%	8. Services (Data Processing)	3.12%
9. Texas Instrs Inc	3.42%	9. Telephone	2.57%
10. America Online Inc.	3.18%	10. Equipment (Semiconductors)	2.44%
Total	62.91%	Total	93.05%

Advisor: State Street Bank
Sponsor: PDR Services LLC
Distributor: ALPS Mutual Funds Inc.

Appendix A — Fund Pages

streetTRACKS Morgan Stanley High Tech 35 Index Fund (Symbol: MTK)

Objective

The streetTRACKS Morgan Stanley High Tech 35 Index Fund seeks investment results that correspond to the price and yield performance, before fees and expenses, of the Morgan Stanley High Tech 35 Index. There is no assurance that the performance of the Morgan Stanley High Tech 35 Index can be fully matched.

Fund Details

Options Traded:	No	Tracking Increment:	$0.01
Expense Ratio:	0.50%	Min Trade Size:	1 Share
Ticker Symbol:	MTK	Marginable:	Yes
Short Selling Allowed:	Yes, uptick exempt		

Distribution History	2000	Quick Facts	
		Net Assets:	$60,137,000
Ordinary Income	$0.00	Shares Outstanding:	1,100,000
Short Term Capital Gains	$0.00	Dividend Yield:	0%
Long Term Capital Gains	$0.00	52 Week High:	$98.25
Return of Capital	$0.00	52 Week Low:	$46.00
Total	$0.00		

Top Ten Holdings		Industry Groups	
1. Xilinx Inc	3.57%	1. Networking & Telecom Equip.	23.11%
2. Texas Instruments Inc	3.16%	2. Internet & PC Software	16.09%
3. Hewlett Packard Co	3.13%	3. Semiconductors	15.67%
4. Micron Technology Inc	3.10%	4. Server & Enterprise Hardware	12.10%
5. EMC Corp	3.06%	5. Enterprise Soft./Tech. Soft.	12.08%
6. International Business Machines	3.04%	6. PC Hardware & Data Storage	7.47%
7. Applied Materials Inc	3.03%	7. Comp. & Business Serv.	6.66%
8. STMicroelectronics NV	2.93%	8. Elect. Manufacturing Serv.	3.79%
9. Intel Corp	2.91%	9. Semiconductor - Capital Equip.	3.03%
10. Sun Microsystems Inc	2.87%	Total	100%

Trustee: State Street Global Advisors
Administrator: State Street Bank & Trust Co.
Distributor: State Street Capital Markets, LLC

iShares Goldman Sachs Technology Index Fund (Symbol: IGM)

Objective

The Goldman Sachs Technology Index Fund seeks investment results that correspond to the price and yield performance, before fees and expenses, of the Goldman Sachs Technology Index. There is no assurance that the performance of the Goldman Sachs Technology Index can be fully matched.

Fund Details			
Options Traded:	No	Trading Increment:	$0.01
Expense Ratio:	0.50%	Min Trade Size:	1 Share
Ticker Symbol:	IGM	Marginable:	Yes
Short Selling Allowed:	Yes, uptick exempt		
Distribution History	**2000**	**Quick Facts**	
		Net Assets:	$64,293,000
Ordinary Income	$0.00	Shares Outstanding:	1,450,000
Short Term Capital Gains	$0.00	Dividend Yield:	0%
Long Term Capital Gains	$0.00	52 Week High:	$56.75
Return of Capital	$0.00	52 Week Low:	$45.46
Total	$0.00	As of Apr. 16, 2001	
Top Ten Holdings		**Industry Groups**	
1. Microsoft Corp	12.07%	1. Computers	23.32%
2. Intel Corp	7.33%	2. Software	23.08%
3. AOL Time Warner Inc	7.19%	3. Semiconductors	20.30%
4. Intl. Business Machines Corp	6.98%	4. Telecommunications	17.59%
5. Cisco Systems Inc	4.71%	5. Media	7.19%
6. Oracle Corp	3.47%	6. Internet	3.94%
7. Dell Computer Corp	2.75%	7. Technology	2.41%
8. EMC Corp	2.66%	8. Commercial Services	1.05%
9. Hewlett-Packard Co.	2.56%	9. Leisure Time	0.25%
10. Texas Instruments Inc.	2.22%	10. Entertainment	0.24%
Total	51.94%	Total	99.37%

Trustee: Barclays Global Fund Advisors.
Administrator: Investors Bank & Trust Company.
Distributor: SEI Investments Distribution Co.

FORTUNE e-50 Index Tracking Stock
(Symbol: FEF)

Objective

The FORTUNE e-50 Index Tracking Stock seeks investment results that correspond to the price and yield performance, before fees and expenses, of the FORTUNE e-50 Index. There is no assurance that the performance of the FORTUNE e-50 Index can be fully matched.

Fund Details

Options Traded:	No	Tracking Increment:	$0.01
Expense Ratio:	0.20%	Min Trade Size:	1 Share
Ticker Symbol:	FEF	Marginable:	Yes
Short Selling Allowed:	Yes, uptick exempt		

Distribution History 2000

This information is currently unavailable

Quick Facts

Net Assets:	$7,300,000
Shares Outstanding:	200,000
Dividend Yield:	0%
52 Week High:	$83.94
52 Week Low:	$29.80

Top Ten Holdings

1. Oracle Corporation	9.21%
2. Microsoft Corp	8.69%
3. Cisco Systems	8.30%
4. America Online Inc	7.43%
5. Intel Corp	6.51%
6. JDS Uniphase Corp	4.45%
7. Juniper Networks Inc.	4.28%
8. Sun Microsystems Inc	3.42%
9. Qwest Communications Int'l	2.99%
10. EMC Corp Mass.	2.96%
Total	58.23%

Industry Groups

1. Hardware	45.02%
2. Software & Services	12.22%
3. eCompanies	10.62%
4. Communications	7.08%
Total	100.00%

Trustee: State Street Global Advisors
Administrator: State Street Bank & Trust Co.
Distributor: State Street Capital Markets, LLC

NASDAQ-100 INDEX TRACKING STOCK (SYMBOL: QQQ)

Objective

The Nasdaq-100 Trust Series I is a pooled investment designed to provide investment results that generally correspond to the price and yield performance of the Nasdaq-100 Index. There is no assurance that the performance of the Nasdaq-100 Index can be fully matched.

Fund Details

Options Traded:	Yes	Trading Increment:	0.01
Expense Ratio:	0.0018	Min Trade Size:	1 Share
Ticker Symbol:	QQQ	Marginable:	Yes
Short Selling Allowed:	Yes, uptick exempt		

Distribution History 2000

Ordinary Income	$0.00
Short Term Capital Gains	$0.00
Long Term Capital Gains	$0.00
Return of Capital	$0.00
Totals	$0.00

Quick Facts

Net Assets:	$23,939,437,500
Shares Outstanding:	580,350,000
Dividend Yield:	0%
52 Week High:	$103.51
52 Week Low:	$33.60

Top Ten Holdings

1. Cisco Systems, Inc.	6.36%
2. Microsoft Corporation	5.15%
3. QUALCOMM Incorporated	4.80%
4. Intel Corporporation	4.59%
5. Oracle Corporation	4.39%
6. Sun Microsystems, Incorporated	2.56%
7. JDS Uniphase Corporation	2.53%
8. VERITAS Software Corporation	2.27%
9. Siebel Systems, Incorporated	2.20%
10. Amgen Incorporated	1.99%
Total	36.83%

Industry Groups

1. Technology	75.33%
2. Health Care	9.78%
3. Communication Services	5.36%
4. Consumer Cyclicals	3.98%
5. Consumer Staples	3.17%
6. Capital Goods	2.14%
7. Basic Materials	0.23%
Total	100%

Trustee: Bank of New York
Administrator: Nasdaq Investment Products Services Inc.
Distributor: ALPS Mutual Funds Inc.

Appendix A — Fund Pages

streetTRACKS Morgan Stanley Internet Index Fund (Symbol: MII)

Objective

The streetTRACKS Morgan Stanley Internet Index Fund seeks investment results that correspond to the price and yield performance, before fees and expenses, of the Morgan Stanley Internet Index. There is no assurance that the performance of the Morgan Stanley Internet Index can be fully matched.

Fund Details

Options Traded:	No	Tracking Increment:	$0.01
Expense Ratio:	0.50%	Min Trade Size:	1 Share
Ticker Symbol:	MII	Marginable:	Yes
Short Selling Allowed:	Yes, uptick exempt		

Distribution History	2000	Quick Facts	
		Net Assets:	$4,450,000
Ordinary Income	$0.00	Shares Outstanding:	250,000
Short Term Capital Gains	$0.00	Dividend Yield:	0%
Long Term Capital Gains	$0.00	52 Week High:	$66.50
Return of Capital	$0.00	52 Week Low:	$13.57
Total	$0.00		

Top Ten Holdings		Industry Groups	
1. Web MD Corp.	5.21%	1. Internet Infrastructure	21.94%
2. CNET Networks Inc.	4.27%	2. Internet Vertical Portals	16.54%
3. I2 Technologies Inc.	4.20%	3. Internet/B2B Software	14.64%
4. Scient Corp	4.20%	4. Internet Portals	13.60%
5. Oracle	3.98%	5. Internet Infrastructure Serv.	10.95%
6. Charles Schwab Corp	3.87%	6. Internet Consulting Serv.	7.75%
7. Juniper Networks Inc.	3.83%	7. Internet Commerce	6.07%
8. EMC Corp. Mass.	3.82%	8. Multi-Sector Internet Sectors	5.11%
9. At Home Corp.	3.79%	9. B2B Commerce	3.40%
10. Sun Microsystems	3.59%		
Total	40.74%	Total	100.00%

Trustee: State Street Global Advisors
Administrator: State Street Bank & Trust Co.
Distributor: State Street Capital Markets, LLC

iShares Dow Jones U.S. Internet Sector Index Fund (Symbol: IYV)

Objective

The Dow Jones U.S. Internet Sector Index Fund seeks investment results that correspond to the price and yield performance, before fees and expenses, of the Dow Jones U.S. Internet Sector Index. There is no assurance that the performance of the Dow Jones U.S. Internet Sector Index can be fully matched.

Annual Performance at NAV

	Inception	2000	2001
IYV	5/15/2000	-51.91%	-47.78%
DJ U.S. Internet Sect Index		-49.80%	-48.94%

Distribution History

	2000	2001
Ordinary Income	$0.00	$0.00
Short Term Capital Gains	$1.64	$0.00
Long Term Capital Gains	$0.00	$0.00
Return of Capital	$0.00	$0.00
Totals	$1.64	$0.00

Fund Details

Expense Ratio:	0.60%
Ticker Symbol:	IYV
Trading Increment:	$0.01
Min. Trade Size:	1 Share
Marginable:	Yes
Options Traded:	No
Short Selling Allowed:	Yes, uptick exempt

Quick Facts

Net Assets:	$17,524,500
Shares Outstanding:	1,050,000
Dividend Yield:	0%
52 Week High:	$82.62
52 Week Low:	$12.32

Information as of: 7/17/01

Top Ten Holdings

1. Verisign Inc	10.46%
2. America Online Inc	10.24%
3. Yahoo! Inc	8.82%
4. BEA Systems Inc	8.43%
5. Ariba Inc	8.30%
6. Exodus Communications	5.94%
7. I2 Technologies Inc	5.42%
8. Check Point Software Tech	5.37%
9. Commerce One Inc	4.15%
10. Inktomi Corp	3.55%
Total	70.67%

Economic Sectors

1. Software	52.82%
2. Consumer Services	27.03%
3. Industrial Services	13.77%
4. Retailers	2.81%
5. Specialty Finance	1.59%
6. Advertising and Media	1.13%
7. Fixed Line Communications	0.60%
8. Industrial Equipment	0.26%
Total	100.01%

Advisor: Barclays Global Fund Advisors
Administrator: PFPC Global Fund Service
Distributor: SEI Investments Distribution Co.

Appendix A — Fund Pages

iShares Dow Jones U.S. Telecommunications Sector Index Fund (Symbol: IYZ)

Objective

The Dow Jones U.S. Telecommunications Sector Index Fund seeks investment results that correspond to the price and yield performance, before fees and expenses, of the Dow Jones U.S. Telecommunications Sector Index. There is no assurance that the performance of the Dow Jones U.S. Telecommunications Sector Index can be fully matched.

Fund Details			
Options Traded:	No	Trading Increment:	$0.01
Expense Ratio:	0.60%	Min Trade Size:	1 Share
Ticker Symbol:	IYZ	Marginable:	Yes
Short Selling Allowed:	Yes, uptick exempt		
Distribution History	**2000**	**Quick Facts**	
		Net Assets:	$52,192,000
Ordinary Income	$0.22	Shares Outstanding:	1,400,000
Short Term Capital Gains	$0.59	Dividend Yield:	0%
Long Term Capital Gains	$0.00	52 Week High:	$63.87
Return of Capital	$0.00	52 Week Low:	$31.50
Total	$0.81	As of Apr. 16, 2001	
Top Ten Holdings		**Industry Groups**	
1. SBC Communications Inc.	16.97%	1. Fixed Line Communications	82.64%
2. Verizon Communications	13.16%	2. Wireless Communications	17.26%
3. AT&T Corp	11.41%		
4. Worldcom Inc	9.12%		
5. BellSouth Corp	7.16%		
6. Qwest Communications Intl.	6.34%		
7. Alltel Corp	2.80%		
8. United States Cellular Corp	2.71%		
9. Telephone & Data System	2.58%		
10. Sprint Corp	2.46%		
Total	74.71%	Total	99.90%

Trustee: Barclays Global Fund Advisors.
Administrator: Investors Bank & Trust Company.
Distributor: SEI Investments Distribution Co.

CYCLICAL/NONCYCLICAL

The Cyclical/Transportation Select Sector SPDR Fund (Symbol: XLY)

Objective

The Cyclical/Transportation Select Sector SPDR Fund is one of nine select sector SPDR funds. Each fund is designed to accumulate and hold a portfolio of common stocks intended to provide investment results that generally correspond to the price performance and yield, before fees and expenses, of a specific select sector index. There is no assurance that the price and yield performance of a specific select sector index can be fully matched. Collectively the nine select sector indexes represent all of the companies in the S&P 500 Index. As of 12/29/00 the Cyclical/Transportation Select Sector SPDR Fund held 66 stocks whose weight within the S&P 500 Index was approximately 6.92 percent.

Fund Details			
Options Traded:	Yes	Trading Increment:	$0.01
Expense Ratio:	0.28%	Min Trade Size:	1 Share
Ticker Symbol:	XLY	Marginable:	Yes
Short Selling Allowed:	Yes, uptick exempt		
Distribution History	**2000**	**Quick Facts**	
Ordinary Income	$0.23	Net Assets:	$116,794,000
Short Term Capital Gains	$0.00	Shares Outstanding:	4,600,000
Long Term Capital Gains	$0.00	Dividend Yield:	1%
Return of Capital	$0.00	52 Week High:	$30.94
Totals	$0.23	52 Week Low:	$21.56
Top Ten Holdings		**Industry Groups**	
1. Wal-Mart Stores Inc.	11.87%	1. Retail - Gen Mer Chain	29.70%
2. Home Depot Inc.	11.64%	2. Retail (Building Supplies)	15.85%
3. Ford Motor Company	10.13%	4. Automobiles	9.04%
4. Target Corporation	5.23%	5. Retail - Dept Stores	5.96%
5. General Motors Corp.	4.92%	3. Retail Speciality - Apparel	4.82%
6. Gap Inc.	4.87%	7. Railroads	4.69%
7. Kohls Corp.	4.22%	8. Airlines	4.49%
8. Southwest Airlines Co.	3.44%	6. Retail Specialty	3.68%
9. Costco Wholesale Corporation	3.42%	9. Leisure Time (Products)	2.93%
10. Lowes Companies Inc.	3.18%	10. Auto Parts & Equipment	2.81%
Total	62.91%	Total	83.99%

Advisor: State Street Bank
Sponsor: PDR Services LLC
Distributor: ALPS Mutual Funds Inc.

Appendix A — Fund Pages

ISHARES DOW JONES U.S. CONSUMER CYCLICAL SECTOR INDEX FUND (SYMBOL: IYC)

Objective

The Dow Jones U.S. Consumer Cyclical Sector Index Fund seeks investment results that correspond to the price and yield performance, before fees and expenses, of the Dow Jones U.S. Consumer Cyclical Sector Index. There is no assurance that the performance of the Dow Jones U.S. Consumer Cyclical Sector Index can be fully matched.

Fund Details			
Options Traded:	No	Trading Increment:	$0.01
Expense Ratio:	0.60%	Min Trade Size:	1 Share
Ticker Symbol:	IYC	Marginable:	Yes
Short Selling Allowed:	Yes, uptick exempt		

Distribution History	2000	Quick Facts	
		Net Assets:	$24,858,000
Ordinary Income	$0.08	Shares Outstanding:	450,000
Short Term Capital Gains	$0.19	Dividend Yield:	0%
Long Term Capital Gains	$0.00	52 Week High:	$63.72
Return of Capital	$0.00	52 Week Low:	$50.96
Total	$0.26	As of Apr. 16, 2001	

Top Ten Holdings		Industry Groups	
1. Wal-Mart Stores Inc	9.04%	1. Retailers	41.13%
2. Home Depot Inc	8.34%	2. Entertainment	26.26%
3. Time Warner Inc	5.83%	3. Advertising and Media	18.49%
4. Walt Disney Co	5.42%	4. Auto Manufacturers & Parts Makers	7.98%
5. Viacom Inc Class B	5.06%	5. Travel	3.38%
6. Ford Motor Co	3.25%	6. Home Construction and Furnishings	1.47%
7. AT&T Corp - Liberty Media Group	2.90%	7. Textile & Apparel	1.28%
8. McDonalds Corp	2.71%		
9. Walgreen Co	2.60%		
10. Comcast Corp Class A	2.43%		
Total	47.58%	Total	99.99%

Trustee: Barclays Global Fund Advisors.
Administrator: Investors Bank & Trust Company.
Distributor: SEI Investments Distribution Co.

iShares Dow Jones US Consumer Non-Cyclical (Symbol: IYK)
Objective

The Dow Jones U.S. Consumer Non-Cyclical Sector Index Fund seeks investment results that correspond to the price and yield performance, before fees and expenses, of the Dow Jones U.S. Consumer Non-Cyclical Sector Index. There is no assurance that the performance of the Dow Jones U.S. Consumer Non-Cyclical Sector Index can be fully matched.

Fund Details

Options Traded:	No	Trading Increment:	$0.01
Expense Ratio:	0.60%	Min Trade Size:	1 Share
Ticker Symbol:	IYK	Marginable:	Yes
Short Selling Allowed:	Yes, uptick exempt		

Distribution History 2000

		Quick Facts	
		Net Assets:	$13,695,500
Ordinary Income	$0.25	Shares Outstanding:	350,000
Short Term Capital Gains	$0.00	Dividend Yield:	1%
Long Term Capital Gains	$0.00	52 Week High:	$44.51
Return of Capital	$0.00	52 Week Low:	$38.32
Total	$0.25	As of Apr. 16, 2001	

Top Ten Holdings

		Industry Groups	
1. America Online Inc	12.99%	1. Food and Beverage Makers	40.16%
2. The Coca-Cola Co	12.25%	2. Consumer Services	21.74%
3. Proctor & Gamble Co	8.66%	3. Household Products	17.65%
4. Pepsico Inc	6.94%	4. Food Retailers & Wholesalers	7.89%
5. Philip Morris Co Inc	6.94%	5. Tobacco	7.76%
6. Anheuser Busch Inc	4.01%	6. Cosmetics	4.63%
7. Yahoo! Inc	3.20%		
8. Kimberly Clark Corp	3.14%		
9. Gillette Co	3.09%		
10. Colgate Palmolive Co	2.63%		
Total	63.87%	Total	99.83%

Trustee: Barclays Global Fund Advisors.
Administrator: Investors Bank & Trust Company.
Distributor: SEI Investments Distribution Co.

REAL ESTATE

iShares Cohen & Steers Realty Majors Index Fund
(Symbol: ICF)

Objective

The iShares Cohen & Steers Realty Majors Index Fund seeks investment results that correspond generally to the price and yield performance, before fees and expenses, of the Cohen & Steers Realty Majors Index. There is no assurance that the performance of the Cohen & Steers Realty Majors Index can be fully matched.

Fund Details			
Options Traded:	No	Trading Increment:	1 cent ($0.01)
Expense Ratio:	0.35%	Min Trade Size:	1 Share
Ticker Symbol:	ICF	Marginable:	Yes
Short Selling Allowed:	Yes, uptick exempt		

Distribution History	2000	Quick Facts	
		Net Assets:	
Ordinary Income		Shares Outstanding:	
Short Term Capital Gains		Dividend Yield:	
Long Term Capital Gains		52 Week High:	
Return of Capital		52 Week Low:	
Total		As of Apr. 16, 2001	

Top Ten Holdings		Industry Groups	
1. Equity Residential Property Trust	8.70%	1. Office	26.90%
2. Equity Office Properties	8.00%	2. Apartment	26.10%
3. Simon Property Group	5.70%	3. Regional Mall	12.60%
4. ProLogis Trust	5.10%	4. Office/Industrial	10.90%
5. Boston Properties	5.10%	5. Industrial	9.50%
6. Apartment Investment & Management	4.90%	6. Shopping Center	5.50%
7. Avalon Bay Communities	4.70%	7. Self Storage	4.50%
8. Vernado Realty Trust	4.70%	8. Health Care	3.00%
9. Spieker Properties	4.60%	9. Manufactured Home	0.09%
10. Public Storage	4.50%		
Total	56.00%	Total	99.09%

Trustee: Barclays Global Fund Advisors.
Administrator: Investors Bank & Trust Company.
Distributor: SEI Investments Distribution Co.

iShares Dow Jones U.S. Real Estate Index Fund (Symbol: IYR)

Objective

The Dow Jones U.S. Real Estate Index Fund seeks investment results that correspond to the price and yield performance, before fees and expenses, of the Dow Jones U.S. Real Estate Index. There is no assurance that the performance of the Dow Jones U.S. Real Estate Index can be fully matched.

Fund Details

Options Traded:	No	Trading Increment:	1 cent ($0.01)
Expense Ratio:	0.60%	Min Trade Size:	1 Share
Ticker Symbol:	IYR	Marginable:	Yes
Short Selling Allowed:	Yes, uptick exempt		

Distribution History 2000

Ordinary Income	$2.42
Short Term Capital Gains	$0.14
Long Term Capital Gains	$0.00
Return of Capital	$0.00
Total	$2.56

Quick Facts

Net Assets:	$37,195,000
Shares Outstanding:	500,000
Dividend Yield:	5%
52 Week High:	$76.90
52 Week Low:	$69.05
As of Apr. 16, 2001	

Top Ten Holdings

1. Equity Office Properties Trust REIT	8.37%
2. Equity Residential Properties Trust REIT	5.64%
3. Prologis Trust	3.72%
4. Spieker Properties Inc REIT	3.59%
5. HomeStore.com Inc	3.45%
6. Vornado Realty Trust REIT	3.07%
7. Avalonbay Communities Inc	3.03%
8. Simon Property Group Inc	2.92%
9. Archstone Communities Trust	2.86%
10. Apartment Investment & Management	2.77%
Total	39.41%

Industry Groups

1. Real Estate	99.38%
Total	99.98%

Trustee: Barclays Global Fund Advisors.
Administrator: Investors Bank & Trust Company.
Distributor: SEI Investments Distribution Co.

Appendix A — Fund Pages

streetTRACKS Wilshire REIT Index Fund (Symbol: RWR)

Objective

The streetTRACKS Wilshire REIT Index Fund seeks investment results that correspond to the price and yield performance, before fees and expenses, of the Wilshire REIT Index. There is no assurance that the performance of the Wilshire REIT Index can be fully matched.

Annual Performance at NAV

	Inception	2001
RWR	4/27/2001	
Wilshire REIT Index		

Distribution History

	2001
Ordinary Income	
Short Term Capital Gains	
Long Term Capital Gains	
Return of Capital	
Totals	

Fund Details

Expense Ratio:	0.25%
Ticker Symbol:	RWR
Trading Increment:	$0.01
Min. Trade Size:	1 Share
Marginable:	Yes
Options Traded:	No
Short Selling Allowed:	Yes, uptick exempt

Quick Facts

Net Assets:	$17,929,500
Shares Outstanding:	150,000
Dividend Yield:	0%
52 Week High:	$120.95
52 Week Low:	$110.55

Information as of: 7/17/01

Top Ten Holdings

1. Equity Office Properties	7.36%
2. Equity Residential Prop.	5.88%
3. Simon Property Group	3.75%
4. Spieker Properties	3.08%
5. Prologis Trust	2.97%
6. Boston Properties, Inc.	2.84%
7. Apartment Inv. & Mgmt. Co.	2.70%
8. Public Storage Props Inc.	2.69%
9. Vornado Realty Trust	2.65%
10. Avalonbay Commun., Inc.	2.62%
Total	35.54%

Economic Sectors

1. Apartment	22.92%
2. Office	22.35%
3. Diversified	13.53%
4. Regional Retail	10.64%
5. Local Retail	8.92%
6. Industrial	8.19%
7. Hotels	6.66%
8. Storage	4.36%
9. Manufactured Homes	1.71%
10. Factory Outlets	0.72%
Total	100.00%

Advisor: State Street Global Advisors
Administrator: State Street Bank & Trust Co.
Distributor: State Street Brokerage Services

ASIA PACIFIC REGION
iShares MSCI Australia Index Fund
(Symbol: EWA)

Objective

The iShares MSCI Australia Index Fund seeks to provide investment results that correspond generally to the price and yield performance of publicly traded securities in the aggregate in the Australian market, as measured by the MSCI Australia Index. There is no assurance that the performance of the MSCI Australia Index can be fully matched.

Annual Performance at NAV

	Inception	1997	1998	1999	2000
EWA	03/18/1996	-10.19%	2.18%	19.24%	-11.82%
MSCI Australia Index		-10.44%	6.07%	17.62%	-9.95%

Distribution History

(Dividends and capital gains declared semi-annually)

	1997	1998	1999	2000
Ordinary Income	$0.15	$0.18	$0.19	$0.24
Short Term Capital Gains	$0.00	$0.00	$0.00	$0.00
Long Term Capital Gains	$0.05	$0.00	$0.00	$0.00
Return of Capital	$0.20	$0.00	$0.06	$0.03
Totals	$0.40	$0.18	$0.25	$0.27

Fund Details

Expense Ratio:	0.0084
Ticker Symbol:	EWA
Trading Increment:	0.01
Min. Trade Size:	1 Share
Marginable:	Yes
Options Traded:	No
Short Selling Allowed:	Yes, uptick exempt

Quick Facts   As of Apr. 18, 2001

Net Assets:	$ 46,800,000
Shares Outstanding:	5,200,000
Dividend Yield:	3%
52 Week High:	10.813
52 Week Low:	8.25

Top Ten Holdings

1. News Corporation	12.52%
2. Telstra	10.55%
3. National Australia Bank Ltd.	9.42%
4. Commonwealth BK of Aust OR	9.01%
5. Broken Hill Propriety Company	5.39%
6. AMP Limited	5.18%
7. Westpac Banking Corporation	4.61%
8. Rio Tinto Limited	2.75%
9. WMC Limited	2.45%
10. Woolworths Limited	2.35%
Total	64.24%

Economic Sectors

1. Banking	23.0%
2. Broadcasting & Publishing	12.5%
3. Telecommuications	10.6%
4. Insurance	6.6%
5. Energy Sources	6.6%
6. Real Estate	6.4%
7. Metals-Non Ferrous	5.8%
8. Business & Public Serv	4.8%
9. Merchandising	4.5%
10. Beverages & Tobacco	4.0%
Total	84.7%

Advisor: Barclays Global Fund Advisors
Administrator: PFPC Global Fund Service
Distributor: SEI Investments Distribution Co.

Appendix A — Fund Pages

iShares MSCI Taiwan Index Fund
(Symbol: EWT)

Objective

The iShares MSCI Taiwan Index Fund seeks to provide investment results that correspond generally to the price and yield performance of publicly traded securities in the aggregate in the Taiwanese market, as measured by the MSCI Taiwan Index. There is no assurance that the performance of the MSCI Taiwan Index can be fully matched.

Annual Performance at NAV

	Inception	2000	2001
EWT	6/23/2000	N/A	17.63%
MSCI Taiwan Index		-44.90%	18.03%

Distribution History

	2000	2001
Ordinary Income	$0.32	$0.00
Short Term Capital Gains	$0.10	$0.00
Long Term Capital Gains	$0.00	$0.00
Return of Capital	$0.44	$0.00
Totals	$0.86	$0.00

Fund Details

Expense Ratio:	0.99%
Ticker Symbol:	EWT
Trading Increment:	$0.01
Min. Trade Size:	1 Share
Marginable:	Yes
Options Traded:	No
Short Selling Allowed:	Yes, uptick exempt

Quick Facts

Net Assets:	$96,492,000
Shares Outstanding:	8,600,000
Dividend Yield:	3%
52 Week High:	$19.44
52 Week Low:	$9.75

Information as of: 4/18/01

Top Ten Holdings

1. Taiwan Semiconductor	18.38%
2. United Micro. Corp.	10.20%
3. Cathay Life Ins	5.34%
4. Hon Hai	5.12%
5. Asustek Computer Inc	4.31%
6. China Dvlp Int'l	3.93%
7. Quanta Computer Inc	3.59%
8. Nan Ya Plastic	3.58%
9. China Steel	2.98%
10. Formosa Plastics Corp.	2.96%
Total	61.99%

Economic Sectors

1. Electronic Components, Instr.	46.70%
2. Banking	12.98%
3. Chemicals	8.59%
4. Data Processing & Reproduc	6.75%
5. Insurance	5.34%
6. Metals-Steel	3.16%
7. Electrical & Electronics	2.51%
8. Building Materials & Comp	2.44%
9. Industrial Components	2.15%
10. Textiles & Apparel	1.62%
Total	91.79%

Advisor: Barclays Global Fund Advisors
Administrator: PFPC Global Fund Service
Distributor: SEI Investments Distribution Co.

iShares MSCI Singapore Index Fund (Symbol: EWS)

Objective

The iShares MSCI Singapore Index Fund seeks to provide investment results that correspond generally to the price and yield performance of publicly traded securities in the aggregate in the Singaporean market, as measured by the MSCI Singapore Index. There is no assurance that the performance of the MSCI Singapore Index can be fully matched.

Annual Performance at NAV

	Inception	1997	1998	1999	2000	2000
EWS	3/18/1996	-43.87%	-5.44%	55.53%	-26.34%	-18.78%
MSCI Singapore Index		-40.46%	-3.59%	60.17%	-27.72%	-18.56%

Distribution History

	1997	1998	1999	2000	2001
Ordinary Income	$0.01	$0.11	$0.14	$0.07	$0.00
Short Term Capital Gains	$0.02	$0.00	$0.00	$0.00	$0.00
Long Term Capital Gains	$0.00	$0.00	$0.00	$0.20	$0.00
Return of Capital	$0.02	$0.01	$0.01	$0.02	$0.00
Totals	$0.05	$0.12	$0.15	$0.29	$0.00

Fund Details

Expense Ratio:	0.84%
Ticker Symbol:	EWS
Trading Increment:	$0.01
Min. Trade Size:	1 Share
Marginable:	Yes
Options Traded:	No
Short Selling Allowed:	Yes, uptick exempt

Quick Facts

Net Assets:	$50,094,000
Shares Outstanding:	9,900,000
Dividend Yield:	1%
52 Week High:	$8.31
52 Week Low:	$5.01
Information as of:	3/30/01

Top Ten Holdings

1. DBS Group Holdings	16.63%
2. Oversea-Chinese Banking	12.59%
3. Singapore Airlines	12.49%
4. Singapore Tele	7.37%
5. Singapore Technologies	5.39%
6. United Overseas Bank	4.25%
7. City Developments Ltd.	4.21%
8. Capital Land	4.18%
9. Singapore Press Hldgs Ltd.	4.04%
10. Chartered Semiconductor	4.00%
Total	75.15%

Economic Sectors

1. Banking	33.47%
2. Real Estate	12.52%
3. Transportation-Airlines	12.49%
4. Electronic Components, Instr	8.73%
5. Multi-Industry	7.60%
6. Telecommunications	7.36%
7. Machinery & Engineering	5.39%
8. Broadcasting & Publishing	4.04%
9. Leisure & Tourism	2.42%
10. Beverages & Tobacco	2.10%
Total	88.26%

Advisor: Barclays Global Fund Advisors
Administrator: PFPC Global Fund Service
Distributor: SEI Investments Distribution Co.

Appendix A — Fund Pages

iShares MSCI Hong Kong Index Fund
(Symbol: EWH)

Objective

The iShares MSCI Hong Kong Index Fund seeks to provide investment results that correspond generally to the price and yield performance of publicly traded securities in the aggregate in the Hong Kong market, as measured by the MSCI Hong Kong Index. There is no assurance that the performance of the MSCI Hong Kong Index can be fully matched.

Annual Performance at NAV

	Inception	1997	1998	1999	2000
EWH	03/18/1996	-26.74%	-9.21%	54.00%	-14.47%
MSCI Hong Kong Index		-23.29%	-2.92%	59.52%	-14.74%

Distribution History

	1997	1998	1999	2000	2001
Ordinary Income	$0.04	$0.18	$0.11	$0.16	$0.00
Short Term Capital Gains	$0.07	$0.01	$0.00	$1.03	$0.00
Long Term Capital Gains	$0.00	$0.04	$0.35	$1.29	$0.00
Return of Capital	$0.03	$0.08	$0.01	$0.04	$0.00
Totals	$0.14	$0.31	$0.47	$2.52	$0.00

Fund Details

Expense Ratio:	0.84%
Ticker Symbol:	EWH
Trading Increment:	$0.01
Min. Trade Size:	1 Share
Marginable:	Yes
Options Traded:	No
Short Selling Allowed:	Yes, uptick exempt

Quick Facts

Net Assets:	$141,556,850
Shares Outstanding:	8,401,000
Dividend Yield:	1%
52 Week High:	$26.25
52 Week Low:	$15.20

Information as of: 3/30/01

Top Ten Holdings

1. Allianz	12.17%
2. Deutsche Telekom	11.36%
3. Siemens	10.45%
4. Muench. Rueckversich.	5.35%
5. Dresdner Bank	5.28%
6. E.On AG	5.09%
7. Deutsche Bank	4.75%
8. DaimlerChrysler	4.62%
9. SAP AG	4.47%
10. Bayer	4.40%
Total	67.94%

Economic Sectors

1. Insurance	17.52%
2. Banking	13.94%
3. Telecommunications	11.36%
4. Electrical & Electronics	10.45%
5. Utilities-Electrical & Gas	8.54%
6. Automobiles	8.29%
7. Chemicals	8.25%
8. Business & Public Services	5.48%
9. Health & Personal Care	4.81%
10. Merchandising	3.09%
Total	91.73%

Advisor: Barclays Global Fund Advisors
Administrator: PFPC Global Fund Service
Distributor: SEI Investments Distribution Co.

iShares MSCI South Korea Index Fund
(Symbol: EWY)

Objective

The iShares MSCI South Korea Index Fund seeks to provide investment results that correspond generally to the price and yield performance of publicly traded securities in the aggregate in the South Korean market, as measured by the MSCI South Korea Index. There is no assurance that the performance of the MSCI South Korea Index can be fully matched.

Annual Performance at NAV

	Inception	2000	2000
EWY	5/12/2000	N/A	1.50%
MSCI South Korea Index		-49.62%	2.01%

Distribution History

	2000	2001
Ordinary Income	$0.00	$0.00
Short Term Capital Gains	$0.00	$0.00
Long Term Capital Gains	$0.00	$0.00
Return of Capital	$0.00	$0.00
Totals	$0.00	$0.00

Fund Details

Expense Ratio:	0.99%
Ticker Symbol:	EWY
Trading Increment:	$0.01
Min. Trade Size:	1 Share
Marginable:	Yes
Options Traded:	No
Short Selling Allowed: Yes, uptick exempt	

Quick Facts

Net Assets:	$16,780,500
Shares Outstanding:	1,350,000
Dividend Yield:	0%
52 Week High:	$22.56
52 Week Low:	$11.01

Information as of: 4/18/01

Top Ten Holdings

1. Samsung Electronics	27.58%
2. SK Telecom	12.36%
3. Korea Elec Power	9.77%
4. Pohang Iron & Steel	5.53%
5. Hyundai MotorCo Ltd	4.25%
6. Kookmin Bank	3.57%
7. Korean Telecom Corp	3.31%
8. Samsung El-Mech Co	2.98%
9. Samsung Display Dev. Ltd	2.39%
10. Ho & Com Bank of Korea	2.21%
Total	73.95%

Economic Sectors

1. Electronic Components, Instr	33.90%
2. Telecommunications	15.67%
3. Multi-Industry	10.14%
4. Banking	8.64%
5. Metals-Steel	5.53%
6. Financial Services	4.89%
7. Automobiles	4.25%
8. Chemicals	2.46%
9. Business & Public Services	1.55%
10. Machinery & Engineering	1.54%
Total	88.57%

Advisor: Barclays Global Fund Advisors
Administrator: PFPC Global Fund Service
Distributor: SEI Investments Distribution Co.

iShares MSCI Japan Index Fund
(Symbol: EWJ)

Objective

The iShares MSCI Japan Index Fund seeks to provide investment results that correspond generally to the price and yield performance of publicly traded securities in the aggregate in the Japanese market, as measured by the MSCI Japan Index. There is no assurance that the performance of the MSCI Japan Index can be fully matched.

Annual Performance at NAV

	Inception	1997	1998	1999	2000	2001
EWJ	3/18/1996	-23.63%	3.53%	57.89%	-28.57%	-8.68%
MSCI Japan Index		23.67%	5.05%	61.53%	-28.16%	-8.44%

Distribution History

	1997	1998	1999	2000	2001
Ordinary Income	$0.00	$0.00	$0.05	$0.00	$0.00
Short Term Capital Gains	$0.00	$0.00	$0.00	$0.03	$0.00
Long Term Capital Gains	$0.01	$0.00	$0.11	$0.39	$0.00
Return of Capital	$0.00	$0.01	$0.01	$0.03	$0.00
Totals	$0.01	$0.01	$0.17	$0.45	$0.00

Fund Details

Expense Ratio:	0.84%
Ticker Symbol:	EWJ
Trading Increment:	$0.01
Min. Trade Size:	1 Share
Marginable:	Yes
Options Traded:	No
Short Selling Allowed:	Yes, uptick exempt

Quick Facts

Net Assets:	$558,910,350
Shares Outstanding:	54,001,000
Dividend Yield:	0%
52 Week High:	$16.19
52 Week Low:	$9.05

Information as of: 3/30/01

Top Ten Holdings

1. Toyota Motor Corp	7.07%
2. Nippon Tel & Telephone	4.40%
3. Sony Corp	3.38%
4. Mizuho Holding	2.59%
5. Bank of Tokyo-Mitsubishi	2.38%
6. Takeda Chemical Ind.	2.38%
7. Matsushita Electric Ind	2.08%
8. Nomura Securities	2.06%
9. Honda Motor Company	2.01%
10. Canon Incorporated	1.79%
Total	30.14%

Economic Sectors

1. Automobiles	10.43%
2. Banking	9.99%
3. Appliances & Household Dur	7.41%
4. Electronic Components, Instr	6.56%
5. Health & Personal Care	6.04%
6. Financial Services	5.20%
7. Electrical & Electronics	5.20%
8. Telecommunications	4.40%
9. Transportation - Road & Rail	3.84%
10. Utilities - Electrical & Gas	3.81%
Total	62.88%

Advisor: Barclays Global Fund Advisors
Administrator: PFPC Global Fund Service
Distributor: SEI Investments Distribution Co.

iShares MSCI Malaysia (Free) Index Fund
(Symbol: EWM)

Objective

The iShares MSCI Malaysia Index Fund seeks to provide investment results that correspond generally to the price and yield performance of publicly traded securities in the aggregate in the Malaysian market, as measured by the MSCI Malaysia Index. There is no assurance that the performance of the MSCI Malaysia Index can be fully matched.

Annual Performance at NAV

	Inception	1997	1998	1999	2000	2001
EWM	3/18/1996	-66.93%	-29.31%	92.98%	-16.09%	-3.77%
MSCI Malaysia Index		-68.11%	-29.49%	114.33%	-15.95%	-3.79%

Distribution History

	1997	1998	1999	2000	2001
Ordinary Income	$0.03	$0.04	$0.00	$0.07	$0.00
Short Term Capital Gains	$0.00	$0.00	$0.00	$0.00	$0.00
Long Term Capital Gains	$0.00	$0.00	$0.00	$0.00	$0.00
Return of Capital	$0.02	$0.20	$0.02	$0.00	$0.00
Totals	$0.05	$0.25	$0.02	$0.07	$0.00

Fund Details

Expense Ratio:	0.84%
Ticker Symbol:	EWM
Trading Increment:	$0.01
Min. Trade Size:	1 Share
Marginable:	Yes
Options Traded:	No
Short Selling Allowed:	Yes, uptick exempt

Quick Facts

Net Assets:	$75,260,250
Shares Outstanding:	18,675,000
Dividend Yield:	2%
52 Week High:	$7.31
52 Week Low:	$3.90

Information as of: 3/30/01

Top Ten Holdings

1. Telkom Malaysia	16.17%
2. Tenga Nasional	15.45%
3. Malayan Banking	12.83%
4. Malaysia Int'l Shipping	5.20%
5. British American Tobacco	4.62%
6. Sime Darby	4.60%
7. Commerce Asset Holdings	3.26%
8. Public Bank (FGN)	3.20%
9. Resorts World	2.98%
10. YTL Corporation	2.28%
Total	70.59%

Economic Sectors

1. Banking	20.87%
2. Telecommunications	16.82%
3. Utilities-Electrical & Gas	15.45%
4. Multi-Industry	7.43%
5. Transportation-Shipping	5.20%
6. Beverages & Tobacco	5.07%
7. Leisure & Tourism	4.88%
8. Misc. Materials & Comm	4.53%
9. Construction & Housing	4.36%
10. Automobiles	2.62%
Total	87.23%

Advisor: Barclays Global Fund Advisors
Administrator: PFPC Global Fund Service
Distributor: SEI Investments Distribution Co.

CANADA

iShares S&P/TSE 60 Index Fund
(Symbol: IKC)

Objective

The S&P/TSE 60 Index Fund seeks investment results that correspond generally to the price and yield performance, before fees and expenses, of the S&P/TSE 60 Index. "TSE" is a trademark of the Toronto Stock Exchange.

Annual Performance at NAV

	Inception	2000	2001
IKC	6/12/2000	-7.81%	-17.35%
S&P/TSE 60 Index		-11.09%	-19.95%

Distribution History

	2000	2001
Ordinary Income	$0.24	$0.04
Short Term Capital Gains	$0.48	$0.00
Long Term Capital Gains	$0.00	$0.00
Return of Capital	$0.00	$0.00
Totals	$0.72	$0.04

Fund Details

Expense Ratio:	0.50%
Ticker Symbol:	IKC
Trading Increment:	$0.01
Min. Trade Size:	1 Share
Marginable:	Yes
Short Selling Allowed:	Yes, uptick exempt

Quick Facts

Net Assets:	$6,817,500
Shares Outstanding:	150,000
Dividend Yield:	1%
52 Week High:	$67.56
52 Week Low:	$43.75
Information as of:	3/30/01

Top Ten Holdings

1. Nortel Networks Corp	11.77%
2. Toronto-Dominion Bank	6.88%
3. BCE Inc	5.57%
4. Canadian Nat'l Railway Co	4.81%
5. Canadian Pacific Limited	4.75%
6. Bank of Nova Scotia	4.62%
7. Canad Imp'l Bank of Comm	4.12%
8. Bombardier Inc Class B	4.11%
9. Manulife Financial Corp	4.03%
10. Magna Int'l Inc Cl A	3.11%
Total	53.77%

Economic Sectors

1. Financial	27.15%
2. Communications	23.93%
3. Industrial	16.58%
4. Basic Materials	11.77%
5. Energy	11.00%
6. Consumer, Cyclical	4.19%
7. Diversified	2.87%
8. Consumer, Non-Cyclical	1.93%
9. Technology	0.25%
10. Diversified	0.25%
Total	99.67%

Advisor: Barclays Global Fund Advisors
Administrator: PFPC Global Fund Service
Distributor: SEI Investments Distribution Co.

iShares MSCI Canada Index Fund
(Symbol: EWC)

Objective

The iShares MSCI Canada Index Fund seeks to provide investment results that correspond generally to the price and yield performance of publicly traded securities in the aggregate in the Canadian market, as measured by the MSCI Canada Index. There is no assurance that the performance of the MSCI Canada Index can be fully matched.

Annual Performance at NAV

Inception: 03/18/1996	1997	1998	1999	2000	2001
EWC	10.91%	-6.47%	46.13%	5.24%	
MSCI Canada Index	12.80%	-6.14%	53.74%	5.34%	

Distribution History

	1997	1998	1999	2000	2001
Ordinary Income	$0.03	$0.15	$0.08	$0.55	$0.00
Short Term Capital Gains	$0.14	$0.00	$0.02	$0.62	$0.00
Long Term Capital Gains	$0.07	$0.55	$0.48	$4.07	$0.00
Return of Capital	$0.00	$0.00	$0.00	$0.12	$0.00
Totals	$0.24	$0.70	$0.58	$5.36	$0.00

Fund Details

Expense Ratio:	0.84%
Ticker Symbol:	EWC
Trading Increment:	$0.01
Min. Trade Size:	1 Share
Marginable:	Yes
Options Traded:	No
Short Selling Allowed: Yes, uptick exempt	

Quick Facts

Net Assets:	$18,240,000
Shares Outstanding:	1,600,000
Dividend Yield:	5%
52 Week High:	$21.81
52 Week Low:	$10.85

As of Apr. 18, 2001

Top Ten Holdings

1. Nortel Networks Corp	19.47%
2. Canadian Imperial Bank	6.05%
3. BCE	5.08%
4. Thomson Corp	4.82%
5. Bank of Nova Scotia	4.08%
6. Bombardier Incorporated	3.94%
7. Celesitca Inc	3.81%
8. Power Corp of Canada	3.58%
9. Manulife Financial Corp.	3.27%
10. Royal Bank of Canada	2.54%
Total	56.64%

Economic Sectors

1. Electrical & Electronics	20.92%
2. Banking	13.56%
3. Energy Sources	10.52%
4. Broadcasting & Publishing	6.08%
5. Insurance	6.04%
6. Financial Services	5.83%
7. Telecommunications	5.08%
8. Metals - Non Ferrous	4.71%
9. Electronic Components, Instr.	4.11%
10. Aeorospace & Military Tech	3.94%
Total	80.79%

Advisor: Barclays Global Fund Advisors
Administrator: PFPC Global Fund Service
Distributor: SEI Investments Distribution Co.

EUROPE

iShares MSCI Belgium Index Fund
(Symbol: EWK)

Objective

The iShares MSCI Belgium Index Fund seeks to provide investment results that correspond generally to the price and yield performance of publicly traded securities in the aggregate in the Belgian market, as measured by the MSCI Belgium Index.

Annual Performance at NAV

	Inception	1997	1998	1999	2000
EWK	03/18/1996	11.84%	51.69%	-14.05%	-16.34%
MSCI Belgium Index		13.55%	67.75%	-14.26%	-16.85%

Distribution History

	1997	1998	1999	2000	2001
Ordinary Income	$0.60	$1.50	$0.01	$0.40	$0.00
Short Term Capital Gains	$0.07	$1.03	$0.29	$0.00	$0.00
Long Term Capital Gains	$0.04	$0.96	$0.89	$0.00	$0.00
Return of Capital	$0.01	$0.10	$0.92	$0.00	$0.00
Totals	$0.73	$3.57	$2.11	$0.40	$0.00

Fund Details

Expense Ratio:	0.84%
Ticker Symbol:	EWK
Trading Increment:	$0.01
Min. Trade Size:	1 Share
Marginable:	Yes
Options Traded:	No
Short Selling Allowed:	Yes, uptick exempt

Quick Facts

Net Assets:	$9,702,000
Shares Outstanding:	840,000
Dividend Yield:	3%
52 Week High:	$14.31
52 Week Low:	$10.20

As of Mar. 30, 2001

Top Ten Holdings

1. Fortis B	23.66%
2. Electrabel	13.13%
3. KBC Bancassur Holding	12.49%
4. Groupe Bruxelles Lambert	5.12%
5. Colruyt	4.93%
6. Delhaize - Le Lion	4.54%
7. UCB	4.52%
8. Algmere Maat voor Nijverh	4.50%
9. Interbrew	4.41%
10. Solvay	4.20%
Total	81.50%

Economic Sectors

1. Insurance	23.66%
2. Banking	13.57%
3. Utilities-Electrc & Gas	13.13%
4. Merchandising	9.46%
5. Multi-Industry	7.64%
6. Health & Personal Care	4.52%
7. Financial Services	4.50%
8. Beverages & Tobacco	4.41%
9. Chemicals	4.20%
10. Rec, Other Cons Goods	4.15%
Total	89.24%

Advisor: Barclays Global Fund Advisors
Administrator: PFPC Global Fund Service
Distributor: SEI Investments Distribution Co.

iShares MSCI Austria Index Fund (Symbol: EWO)

Objective

The iShares MSCI Austria Index Fund seeks to provide investment results that correspond generally to the price and yield performance of publicly traded securities in the aggregate in the Austrian market, as measured by the MSCI Austria Index. There is no assurance that the performance of the MSCI Austria Index can be fully matched.

Annual Performance at NAV

	Inception	1997	1998	1999	2000
EWO	03/18/1996	1.05%	-1.83%	-10.36%	-10.57%
MSCI Austria Index		-1.57%	0.35%	-9.11%	-11.96%

Distribution History
(Dividends and capital gains declared semiannually)

	1997	1998	1999	2000
Ordinary Income	$0.02	$0.18	$0.08	$0.00
Short Term Capital Gains	$0.61	$0.61	$0.02	$0.00
Long Term Capital Gains	$0.00	$0.50	$0.00	$0.00
Return of Capital	$0.00	$0.00	$0.00	$0.00
Totals	$0.02	$0.79	$0.10	$0.04

Fund Details

Expense Ratio:	0.84%
Ticker Symbol:	EWO
Trading Increment:	$0.01
Min. Trade Size:	1 Share
Marginable:	Yes
Options Traded:	No
Short Selling Allowed:	Yes, uptick exempt

Quick Facts (As of Apr. 18, 2000)

Net Assets:	$11,312,000
Shares Outstanding:	1,400,000
Dividend Yield:	1%
52 Week High:	$8.70
52 Week Low:	$6.75

Top Ten Holdings

1. Bank Austria	23.98%
2. Oester Elerktriz	15.28%
3. OMV	9.41%
4. Erste Bank Der Oesterreichischen	4.99%
5. Austria Tabakwerke	4.89%
6. Boehler-Uddeholm	4.67%
7. Generali Holding Vienna	4.64%
8. Flughafen Wein	4.41%
9. RHI Ag	4.02%
10. Weinerberger Baustoff	3.96%
Total	80.26%

Economic Sectors

1. Banking	23.98%
2. Utilities — Electric & Gas	15.28%
3. Metals — Steel	9.67%
4. Energy Sources	9.41%
5. Beverages & Tobacco	8.56%
6. Misc. Materials & Comm	5.77%
7. Machinery & Engineering	5.14%
8. Insurance	4.64%
9. Business & Public Svcs	4.41%
10. Bldg Mat/Compon	3.96%
Total	90.81%

Advisor: Barclays Global Fund Advisors
Administrator: PFPC Global Fund Service
Distributor: SEI Investments Distribution Co.

Appendix A — Fund Pages

iSHARES MSCI SPAIN INDEX FUND
(SYMBOL: EWP)

Objective

The iShares MSCI Spain Index Fund seeks to provide investment results that correspond generally to the price and yield performance of publicly traded securities in the aggregate in the Spanish market, as measured by the MSCI Spain Index. There is no assurance that the performance of the MSCI Spain Index can be fully matched.

Annual Performance at NAV

	Inception	1997	1998	1999	2000	2001
EWP	3/18/1996	23.90%	51.30%	-2.12%	-13.62%	-0.94%
MSCI Spain Index		25.41%	49.90%	4.83%	-15.86%	-2.48%

Distribution History

	1997	1998	1999	2000	2001
Ordinary Income	$0.17	$0.13	$0.09	$0.14	$0.00
Short Term Capital Gains	$0.32	$0.07	$0.07	$0.00	$0.00
Long Term Capital Gains	$0.53	$0.86	$0.89	$0.80	$0.00
Return of Capital	$0.09	$0.06	$0.04	$0.01	$0.00
Totals	$1.11	$1.12	$1.09	$0.95	$0.00

Fund Details

Expense Ratio:	0.84%
Ticker Symbol:	EWP
Trading Increment:	$0.01
Min. Trade Size:	1 Share
Marginable:	Yes
Options Traded:	No
Short Selling Allowed:	Yes, uptick exempt

Quick Facts

Net Assets:	$29,707,500
Shares Outstanding:	1,275,000
Dividend Yield:	1%
52 Week High:	$28.19
52 Week Low:	$20.75

Information as of: 4/18/01

Top Ten Holdings

1. Telefonica	25.05%
2. Banco Bilbao Vizcaya	12.48%
3. Banco Santander Cent His	11.80%
4. Repsol	5.13%
5. Endesa	4.73%
6. Iberdrola	4.72%
7. Union Electronica Fenosa	4.68%
8. Gas Natural SDG	4.29%
9. Acerinox	3.41%
10. Autopistas Concesi Esp	2.61%
Total	78.90%

Economic Sectors

1. Telecommunications	25.05%
2. Banking	24.29%
3. Utilities-Electrical & Gas	18.42%
4. Energy Sources	5.13%
5. Business & Public Services	5.01%
6. Construction & Housing	3.61%
7. Metals-Steel	3.41%
8. Health & Personal Care	2.32%
9. Leisure & Tourism	2.19%
10. Machinery & Engineering	1.73%
Total	91.16%

Advisor: Barclays Global Fund Advisors
Administrator: PFPC Global Fund Service
Distributor: SEI Investments Distribution Co.

iShares MSCI Germany Index Fund (Symbol: EWG)

Objective

The iShares MSCI Germany Index Fund seeks to provide investment results that correspond generally to the price and yield performance of publicly traded securities in the aggregate in the German market, as measured by the MSCI Germany Index. There is no assurance that the performance of the MSCI Germany Index can be fully matched.

Annual Performance at NAV

	Inception	1997	1998	1999	2000
EWG	03/18/1996	22.75%	28.28%	20.87%	-15.97%
MSCI Germany Index		24.57%	29.43%	20.04%	-15.59%

Distribution History

	1997	1998	1999	2000	2001
Ordinary Income	$0.04	$0.18	$0.11	$0.16	$0.00
Short Term Capital Gains	$0.07	$0.01	$0.00	$1.03	$0.00
Long Term Capital Gains	$0.00	$0.04	$0.35	$1.29	$0.00
Return of Capital	$0.03	$0.08	$0.01	$0.04	$0.00
Totals	$0.14	$0.31	$0.47	$2.52	$0.00

Fund Details

Expense Ratio:	0.84%
Ticker Symbol:	EWG
Trading Increment:	$0.01
Min. Trade Size:	1 Share
Marginable:	Yes
Options Traded:	No
Short Selling Allowed:	Yes, uptick exempt

Quick Facts

Net Assets:	$141,556,850
Shares Outstanding:	8,401,000
Dividend Yield:	1%
52 Week High:	$26.25
52 Week Low:	$15.20

Information as of: 3/30/01

Top Ten Holdings

1. Allianz	12.17%
2. Deutsche Telekom	11.36%
3. Siemens	10.45%
4. Muench. Rueckversich.	5.35%
5. Dresdner Bank	5.28%
6. E.On AG	5.09%
7. Deutsche Bank	4.75%
8. DaimlerChrysler	4.62%
9. SAP AG	4.47%
10. Bayer	4.40%
Total	67.94%

Economic Sectors

1. Insurance	17.52%
2. Banking	13.94%
3. Telecommunications	11.36%
4. Electrical & Electronics	10.45%
5. Utilities-Electrical & Gas	8.54%
6. Automobiles	8.29%
7. Chemicals	8.25%
8. Business & Public Services	5.48%
9. Health & Personal Care	4.81%
10. Merchandising	3.09%
Total	91.73%

Advisor: Barclays Global Fund Advisors
Administrator: PFPC Global Fund Service
Distributor: SEI Investments Distribution Co.

Appendix A — Fund Pages

iShares MSCI EMU Index Fund
(Symbol: EZU)

Objective

The iShares MSCI EMU Index Fund seeks to provide investment results that correspond generally to the price and yield performance of publicly traded securities in the aggregate in the European Monetary Union (EMU) markets, as measured by the MSCI EMU Index. There is no assurance that the performance of the MSCI EMU Index can be fully matched.

Annual Performance at NAV

	Inception	2000	2001
EZU	07/14/2000	N/A	-16.29%
MSCI EMU Index		-8.72%	-16.22%

Distribution History

(Dividends and capital gains declared semi-annually)

	2000	2001
Ordinary Income	$0.00	$0.08
Short Term Capital Gains	$0.00	$0.02
Long Term Capital Gains	$0.00	$0.48
Return of Capital	$0.00	$0.00
Totals	$0.00	$0.58

Fund Details

Expense Ratio:	0.84%
Ticker Symbol:	EZU
Trading Increment:	$0.01
Min. Trade Size:	1 Share
Marginable:	Yes
Options Traded:	0.84%
Short Selling Allowed:	Yes, uptick exempt

Quick Facts

Net Assets:	$50,360,000
Shares Outstanding:	800,000
Dividend Yield:	0%
52 Week High:	$79.19
52 Week Low:	$56.84

As of Apr. 18, 2001

Top Ten Holdings

1. Royal Dutch Pet./Shell Grp	4.07%
2. Nokia Corporation	3.85%
3. Total - Class B	3.43%
4. Vivendi	2.76%
5. Deutsche Telekom	2.55%
6. Telefonica SA	2.49%
7. Allianz	2.36%
8. ING Groep	2.24%
9. Siemens	2.17%
10. Adventis SA	2.15%
Total	28.07%

Economic Sectors

1. Banking	13.34%
2. Telecommunications	11.46%
3. Energy Sources	10.13%
4. Insurance	10.04%
5. Electrical & Electronics	7.89%
6. Business & Public Services	7.36%
7. Health & Personal Care	6.89%
8. Utilities - Electrical & Gas	5.25%
9. Merchandising	4.24%
10. Chemicals	3.02%
Total	79.62%

Advisor: Barclays Global Fund Advisors
Administrator: PFPC Global Fund Service
Distributor: SEI Investments Distribution Co.

iShares S&P Europe 350 Index Fund (Symbol: IEV)

Objective

The S&P Europe 350 Index Fund seeks investment results that correspond generally to the price and yield performance, before fees and expenses, of the S&P Europe 350 Index™. There is no assurance that the performance of the S&P Europe 350 Index can be fully matched.

Annual Performance at NAV

	Inception	2000	2001
IEV	7/25/2000	-5.09%	-15.82%
S&P Europe 350 Index		-1.71%	-15.88%

Distribution History

	2000	2001
Ordinary Income	$0.19	$0.09
Short Term Capital Gains	$0.00	$0.00
Long Term Capital Gains	$0.00	$0.00
Return of Capital	$0.00	$0.00
Totals	$0.19	$0.09

Fund Details

Expense Ratio:	0.60%
Ticker Symbol:	IEV
Trading Increment:	$0.01
Min. Trade Size:	1 Share
Marginable:	Yes
Options Traded:	No
Short Selling Allowed:	Yes, uptick exempt

Quick Facts

Net Assets:	$159,985,000
Shares Outstanding:	2,450,000
Dividend Yield:	0%
52 Week High:	$80.75
52 Week Low:	$59.02

Information as of: 3/30/01

Top Ten Holdings

1. BP Amoco	3.91%
2. Vodafone Group PLC	3.70%
3. GlaxoSmithKline PLC	3.41%
4. Novartis AG - Reg Shares	2.47%
5. HSBC Hldg.	2.32%
6. Nokia OJY	2.20%
7. Royal Dutch Petro Co	2.17%
8. Total SA - Series B	2.09%
9. Zeneca Group PLC	1.77%
10. Nestle SA	1.65%
Total	25.69%

Economic Sectors

1. Financial	27.93%
2. Consumer, Noncyclical	20.59%
3. Communications	16.67%
4. Energy	11.25%
5. Industrial	6.50%
6. Consumer, Cyclical	6.37%
7. Utilities	4.24%
8. Basic Materials	3.39%
9. Technology	2.13%
10. Diversified	0.39%
Total	99.46%

Advisor: Barclays Global Fund Advisors
Administrator: PFPC Global Fund Service
Distributor: SEI Investments Distribution Co.

Appendix A — Fund Pages

iShares MSCI Italy Index Fund
(Symbol: EWI)

Objective

The iShares MSCI Italy Index Fund seeks to provide investment results that correspond generally to the price and yield performance of publicly traded securities in the aggregate in the Italian market, as measured by the MSCI Italy Index. There is no assurance that the performance of the MSCI Italy Index can be fully matched.

Annual Performance at NAV

	Inception	1997	1998	1999	2000	2001
EWI	3/18/1996	35.77%	50.24%	0.53%	-1.17%	-14.17%
MSCI Italy Index		35.48%	52.52%	-0.26%	-1.33%	-14.24%

Distribution History

	1997	1998	1999	2000	2001
Ordinary Income	$0.38	$1.17	$0.07	$0.12	$0.00
Short Term Capital Gains	$0.00	$0.35	$0.12	$0.00	$0.00
Long Term Capital Gains	$0.00	$5.02	$1.64	$1.80	$0.00
Return of Capital	$0.00	$0.00	$0.26	$0.32	$0.00
Totals	$0.38	$6.54	$2.08	$2.24	$0.00

Fund Details

Expense Ratio:	0.84%
Ticker Symbol:	EWI
Trading Increment:	$0.01
Min. Trade Size:	1 Share
Marginable:	Yes
Options Traded:	No
Short Selling Allowed:	Yes, uptick exempt

Quick Facts

Net Assets:	$38,025,000
Shares Outstanding:	1,950,000
Dividend Yield:	1%
52 Week High:	$27
52 Week Low:	$17.80

Information as of: 3/30/01

Top Ten Holdings

1. Telecom Italia Mobile SpA	14.35%
2. ENI SpA	14.23%
3. Telecom Italia SpA	11.62%
4. Assicurazioni Generali SpA	10.92%
5. Enel SpA	4.67%
6. Unicredito Italiano SpA	4.47%
7. Banco Intesa SpA	4.41%
8. Instituto Banc san Paolo	4.36%
9. Fiat SpA	3.32%
10. Mediaset SpA	2.93%
Total	75.28%

Economic Sectors

1. Telecommunications	25.97%
2. Banking	17.77%
3. Insurance	14.26%
4. Energy Sources	14.23%
5. Utilities-Electrical & Gas	6.39%
6. Broadcasting & Publishing	4.12%
7. Automobiles	3.32%
8. Business & Public Services	2.55%
9. Financial Services	2.44%
10. Electrical & Electronics	2.03%
Total	93.08%

Advisor: Barclays Global Fund Advisors
Administrator: PFPC Global Fund Service
Distributor: SEI Investments Distribution Co.

iShares MSCI Netherlands Index Fund
(Symbol: EWN)

Objective

The iShares MSCI Netherlands Index Fund seeks to provide investment results that correspond generally to the price and yield performance of publicly traded securities in the aggregate in the Dutch market, as measured by the MSCI Netherlands Index. There is no assurance that the performance of the MSCI Netherlands Index can be fully matched.

Annual Performance at NAV

	Inception	1997	1998	1999	2000	2001
EWN	3/18/1996	20.11%	24.09%	4.54%	-7.80%	-14.41%
MSCI Netherlands Index		23.77%	23.23%	6.88%	-4.09%	-15.52%

Distribution History

	1997	1998	1999	2000	2001
Ordinary Income	$0.10	$0.21	$0.39	$0.08	$0.00
Short Term Capital Gains	$0.04	$0.00	$0.18	$0.00	$0.00
Long Term Capital Gains	$0.81	$1.69	$1.12	$0.11	$0.00
Return of Capital	$0.02	$0.07	$0.08	$0.04	$0.00
Totals	$0.96	$1.97	$1.77	$0.23	$0.00

Fund Details		Quick Facts	
Expense Ratio:	0.84%	Net Assets:	$31,875,910
Ticker Symbol:	EWN	Shares Outstanding:	1,601,000
Trading Increment:	$0.01	Dividend Yield:	0%
Min. Trade Size:	1 Share	52 Week High:	$25.50
Marginable:	Yes	52 Week Low:	$18.10
Options Traded:	No		
Short Selling Allowed: Yes, uptick exempt		Information as of: 3/30/01	

Top Ten Holdings		Economic Sectors	
1. Royal Dutch/Shell Group	24.64%	1. Energy Sources	24.64%
2. Ing Groep	11.57%	2. Financial Services Insurance	11.57%
3. Aegon N.V.	7.40%	3. Insurance	9.47%
4. Philips Electronics	5.89%	4. Merchandising	6.12%
5. Koninklijke Ahold	5.25%	5. Appliances & Household Dur	5.89%
6. Heineken	5.17%	6. Beverages & Tobacco	5.17%
7. Unilever	4.79%	7. Food & Household Products	5.14%
8. ABN Amro Holding	4.45%	8. Business & Public Services	5.00%
9. Akzo Nobel	4.05%	9. Chemicals	4.94%
10. TNT Post Group	3.39%	10. Broadcasting & Publishing	4.59%
Total	76.60%	Total	82.53%

Advisor: Barclays Global Fund Advisors
Administrator: PFPC Global Fund Service
Distributor: SEI Investments Distribution Co.

iShares MSCI Switzerland Index Fund
(Symbol: EWL)

Objective

The iShares MSCI Switzerland Index Fund seeks to provide investment results that correspond generally to the price and yield performance of publicly traded securities in the aggregate in the Swiss market, as measured by the MSCI Switzerland Index. There is no assurance that the performance of the MSCI Switzerland Index can be fully matched.

Annual Performance at NAV

	Inception	1997	1998	1999	2000	2001
EWL	3/18/1996	35.23%	18.27%	-3.25%	-5.87%	-16.48%
MSCI Switzerland Index		44.25%	23.53%	-7.02%	-5.85%	-17.58%

Distribution History

	1997	1998	1999	2000	2001
Ordinary Income	$0.00	$0.01	$0.07	$0.04	$0.00
Short Term Capital Gains	$0.10	$0.29	$0.14	$0.10	$0.00
Long Term Capital Gains	$0.47	$0.93	$0.18	$0.01	$0.00
Return of Capital	$0.00	$0.03	$0.01	$0.01	$0.00
Totals	$0.57	$1.25	$0.39	$0.16	$0.00

Fund Details

Expense Ratio:	0.84%
Ticker Symbol:	EWL
Trading Increment:	$0.01
Min. Trade Size:	1 Share
Marginable:	Yes
Options Traded:	No
Short Selling Allowed:	Yes, uptick exempt

Quick Facts

Net Assets:	$37,420,500
Shares Outstanding:	2,626,000
Dividend Yield:	0%
52 Week High:	$17.75
52 Week Low:	$12.81

Information as of: 4/18/01

Top Ten Holdings

1. Novartis	20.94%
2. Nestle	10.80%
3. UBS	9.77%
4. Roche Holding AG – Gen	8.71%
5. Credit Suisse Group	5.10%
6. Schweizerische Rueckvers	4.86%
7. ABB Ltd	4.08%
8. Zurich Allied	3.50%
9. Julius Baer Holding	3.09%
10. Banque Cantonale Vaudoi	3.01%
Total	74.01%

Economic Sectors

1. Health & Personal Care	30.58%
2. Chemicals	25.16%
3. Banking	19.07%
4. Food & Household Products	10.80%
5. Insurance	8.36%
6. Electrical & Electronics	5.26%
7. Machinery & Engineering	4.50%
8. Building Mat & Components	3.26%
9. Financial Services	3.09%
10. Telecommunication	2.81%
Total	90.74%

Advisor: Barclays Global Fund Advisors
Administrator: PFPC Global Fund Service
Distributor: SEI Investments Distribution Co.

iShares MSCI France Index Fund (Symbol: EWQ)

Objective

The iShares MSCI France Index Fund seeks to provide investment results that correspond generally to the price and yield performance of publicly traded securities in the aggregate in the French market, as measured by the MSCI France Index. There is no assurance that the performance of the MSCI France Index can be fully matched.

Annual Performance at NAV

	Inception	1997	1998	1999	2000
EWQ	03/18/1996	11.47%	40.78%	29.97%	-5.09%
MSCI France Index		11.94%	41.54%	29.27%	-4.31%

Distribution History

	1997	1998	1999	2000	2001
Ordinary Income	$0.15	$0.21	$0.12	$0.11	$0.00
Short Term Capital Gains	$0.20	$0.13	$0.19	$0.21	$0.00
Long Term Capital Gains	$0.03	$0.11	$0.59	$0.89	$0.00
Return of Capital	$0.00	$0.07	$0.08	$0.03	$0.00
Totals	$0.38	$0.40	$0.81	$1.23	$0.00

Fund Details

Expense Ratio:	0.84%
Ticker Symbol:	EWQ
Trading Increment:	$0.01
Min. Trade Size:	1 Share
Marginable:	Yes
Options Traded:	No
Short Selling Allowed:	Yes, uptick exempt

Quick Facts

Net Assets:	$67,541,100
Shares Outstanding:	3,201,000
Dividend Yield:	1%
52 Week High:	$29.94
52 Week Low:	$18.90

Top Ten Holdings

1. Total-Class B	11.84%
2. Vivendi	7.51%
3. France Telecom	7.05%
4. Aventis	7.00%
5. L'OREAL	5.37%
6. AXA	5.17%
7. Sanofi-Synthelabo	4.74%
8. Banque National Paris	4.56%
9. Carrefour	4.37%
10. Alcatel	4.17%
Total	61.78%

Economic Sectors

1. Health & Personal Care	17.52%
2. Business & Public Services	14.60%
3. Energy Sources	11.84%
4. Banking	7.73%
5. Merchandising	7.65%
6. Telecommunications	7.05%
7. Electrical & Electronics	5.63%
8. Insurance	5.17%
9. Electrical Compon. & Instr	3.49%
10. Appliances	3.08%
Total	83.76%

Advisor: Barclays Global Fund Advisors
Administrator: PFPC Global Fund Service
Distributor: SEI Investments Distribution Co.

Appendix A — Fund Pages

iShares MSCI United Kingdom Index Fund (Symbol: EWU)

Objective

The iShares MSCI United Kingdom Index Fund seeks to provide investment results that correspond generally to the price and yield performance of publicly traded securities in the aggregate in the British market, as measured by the MSCI United Kingdom Index. There is no assurance that the performance of the MSCI United Kingdom Index can be fully matched.

Annual Performance at NAV

	Inception	1997	1998	1999	2000	2001
EWU	3/18/1996	20.85%	18.42%	12.14%	-11.75%	-12.56%
MSCI UK Index		22.62%	17.80%	12.45%	-11.53%	-12.30%

Distribution History

	1997	1998	1999	2000	2001
Ordinary Income	$0.33	$0.35	$0.33	$0.25	$0.00
Short Term Capital Gains	$0.00	$0.02	$0.03	$0.09	$0.00
Long Term Capital Gains	$0.17	$0.16	$0.92	$0.56	$0.00
Return of Capital	$0.10	$0.07	$0.09	$0.05	$0.00
Totals	$0.60	$0.60	$1.37	$0.95	$0.00

Fund Details

Expense Ratio:	0.84%
Ticker Symbol:	EWU
Trading Increment:	$0.01
Min. Trade Size:	1 Share
Marginable:	Yes
Options Traded:	No
Short Selling Allowed:	Yes, uptick exempt

Top Ten Holdings

1. Glaxosmithkline	8.84%
2. Vodafone Airtouch	8.83%
3. BP Amoco	7.47%
4. Astrazeneca	5.40%
5. HSBC Holdings	4.38%
6. Royal Bank of Scotland Gr	3.82%
7. Lloyds TSB Group	3.58%
8. Barclays	3.20%
9. British Telecommunicaitons	3.17%
10. CGNU	1.85%
Total	50.54%

Advisor: Barclays Global Fund Advisors
Administrator: PFPC Global Fund Service
Distributor: SEI Investments Distribution Co.

Quick Facts

Net Assets:	$122,984,180
Shares Outstanding:	7,601,000
Dividend Yield:	2%
52 Week High:	$20.25
52 Week Low:	$14.55

Information as of: 4/18/01

Economic Sectors

1. Banking	18.17%
2. Health & Personal Care	15.23%
3. Telecommunications	12.01%
4. Energy Sources	7.47%
5. Business & Public Services	6.77%
6. Utilities-Electrical & Gas	5.12%
7. Merchandising	4.04%
8. Insurance	3.79%
9. Beverages & Tobacco	3.79%
10. Broadcasting & Publishing	3.46%
Total	79.85%

iShares MSCI Sweden Index Fund (Symbol: EWD)

Objective

The iShares MSCI Sweden Index Fund seeks to provide investment results that correspond generally to the price and yield performance of publicly traded securities in the aggregate in the Swedish market, as measured by the MSCI Sweden Index. There is no assurance that the performance of the MSCI Sweden Index can be fully matched.

Annual Performance at NAV

	Inception	1997	1998	1999	2000	2001
EWD	3/18/1996	11.00%	11.06%	63.93%	-23.74%	-26.97%
MSCI Sweden Index		12.92%	13.96%	79.74%	-21.29%	-29.89%

Distribution History

	1997	1998	1999	2000	2001
Ordinary Income	$0.00	$0.09	$0.10	$0.14	$0.00
Short Term Capital Gains	$0.18	$0.01	$0.00	$0.53	$0.00
Long Term Capital Gains	$0.62	$0.89	$1.58	$4.69	$0.00
Return of Capital	$0.00	$0.02	$0.01	$0.04	$0.00
Totals	$0.80	$1.01	$1.69	$5.39	$0.00

Fund Details

Expense Ratio:	0.84%
Ticker Symbol:	EWD
Trading Increment:	$0.01
Min. Trade Size:	1 Share
Marginable:	Yes
Options Traded:	No
Short Selling Allowed:	Yes, uptick exempt

Quick Facts

Net Assets:	$11,426,250
Shares Outstanding:	825,000
Dividend Yield:	1%
52 Week High:	$34.31
52 Week Low:	$12.50

Information as of: 4/18/01

Top Ten Holdings

1. Ericsson LM	23.69%
2. Nordea	10.73%
3. Hennes & Mauritz AB	6.78%
4. Svenska Handelsbanken	6.19%
5. Volvo AB	5.02%
6. Telia AB	4.87%
7. Skandia Forsakrings AB	4.59%
8. Svenska Cellulosa	4.13%
9. Assa Abloy AB	4.09%
10. Tele2 AB	3.89%
Total	73.98%

Economic Sectors

1. Electrical & Electronics	23.69%
2. Banking	20.59%
3. Machinery & Engineering	9.36%
4. Telecommunications	8.76%
5. Merchandising	6.78%
6. Business & Public Services	5.34%
7. Insurance	4.59%
8. Forest Pdts & Paper	4.13%
9. Building Mats & Components	4.09%
10. Appliances & House Dur	2.93%
Total	90.26%

Advisor: Barclays Global Fund Advisors
Administrator: PFPC Global Fund Service
Distributor: SEI Investments Distribution Co.

LATIN AMERICA

iShares MSCI Brazil Index Fund
(Symbol: EWZ)

Objective

The iShares MSCI Brazil Index Fund seeks to provide investment results that correspond generally to the price and yield performance of publicly traded securities in the aggregate in the Brazilian market, as measured by the MSCI Brazil Index. There is no assurance that the performance of the MSCI Brazil Index can be fully matched.

Annual Performance at NAV

	Inception	2000	2001
EWZ	7/14/2000	N/A	-9.46%
MSCI Brazil Index		-11.37%	-9.90%

Distribution History

	2000	2001
Ordinary Income	$0.02	$0.00
Short Term Capital Gains	$0.34	$0.00
Long Term Capital Gains	$0.00	$0.00
Return of Capital	$0.00	$0.00
Totals	$0.36	$0.00

Fund Details

Expense Ratio:	0.99%
Ticker Symbol:	EWZ
Trading Increment:	$0.01
Min. Trade Size:	1 Share
Marginable:	Yes
Options Traded:	No
Short Selling Allowed:	Yes, uptick exempt

Quick Facts

Net Assets:	$15,277,500
Shares Outstanding:	1,050,000
Dividend Yield:	0%
52 Week High:	$20.37
52 Week Low:	$13.97

Information as of: 3/30/01

Top Ten Holdings

1. Petroleo Brasileiro	20.42%
2. Centrais Electr Brasileria	10.74%
3. Com Vale do Rio Doce Pfd	10.54%
4. Cia De Bebi Das Americas	8.64%
5. Tele Norte Leste Part. PN	6.61%
6. Banco Itau Pfd	4.40%
7. Banco Bradesco PN	3.89%
8. Emp Brasileria de Aeronut	3.77%
9. Comp Brasileira de Distrib	3.56%
10. Tele Centro Particip Pfd	3.12%
Total	75.69%

Economic Sectors

1. Energy Sources	21.00%
2. Telecommunications	18.55%
3. Utilities -Electrical & Gas	13.75%
4. Metals - Steel	13.22%
5. Beverages & Tobacco	10.65%
6. Banking	10.15%
7. Aerospace & Military Techn	3.77%
8. Merchandising	3.68%
9. Forest Products & Paper	2.51%
10. Chemicals	0.73%
Total	98.01%

Advisor: Barclays Global Fund Advisors
Administrator: PFPC Global Fund Service
Distributor: SEI Investments Distribution Co.

iShares MSCI Mexico Index Fund (Symbol: EWW)

Objective

The iShares MSCI Mexico Index Fund seeks to provide investment results that correspond generally to the price and yield performance of publicly traded securities in the aggregate in the Mexican market, as measured by the MSCI Mexico Index. There is no assurance that the performance of the MSCI Mexic Index can be fully matched.

Annual Performance at NAV

	Inception	1997	1998	1999	2001
EWW	3/18/1996	48.53%	-35.00%	76.12%	3.34%
MSCI Mexico Index		53.29%	-29.49%	80.07%	3.77%

Distribution History

	1997	1998	1999	2001
Ordinary Income	$0.03	$0.12	$0.03	$0.00
Short Term Capital Gains	$0.31	$0.00	$0.00	$0.00
Long Term Capital Gains	$0.13	$0.30	$0.00	$0.00
Return of Capital	$0.05	$0.00	$0.06	$0.00
Totals	$0.52	$0.42	$0.09	$0.00

Fund Details

Expense Ratio:	0.84%
Ticker Symbol:	EWW
Trading Increment:	$0.01
Min. Trade Size:	1 Share
Marginable:	Yes
Options Traded:	No
Short Selling Allowed:	Yes, uptick exempt

Quick Facts

Net Assets:	$42,560,000
Shares Outstanding:	2,800,000
Dividend Yield:	0%
52 Week High:	$19.19
52 Week Low:	$12.37

Information as of: 3/30/01

Top Ten Holdings

1. Telefonos de Mexico Ser L	21.39%
2. America Movil SA	10.70%
3. Walmart de Mexico	7.55%
4. Grupo Fin Banamex Acci	7.25%
5. Fomento Economico Mexi	5.37%
6. Cemex	5.24%
7. Grupo Fin BBVA Bancomer	5.14%
8. Kimberly-Clark	4.90%
9. Grupo Modelo - Series C	4.65%
10. Grupo Televisa Ser CPO	3.64%
Total	75.83%

Economic Sectors

1. Telecommunications	32.09%
2. Beverages & Tobacco	14.11%
3. Banking	12.40%
4. Merchandising	11.90%
5. Multi-Industry	7.30%
6. Building Mat & Components	5.24%
7. Health & Personal Care	4.90%
8. Broadcasting & Publishing	3.64%
9. Food & Household Products	3.43%
10. Metals-Non Ferrous	2.95%
Total	97.96%

Advisor: Barclays Global Fund Advisors
Administrator: PFPC Global Fund Service
Distributor: SEI Investments Distribution Co.

Appendix A — Fund Pages

HOLDRS

HOLDRS, developed by Merrill Lynch, are highly liquid and efficient securities that provide diversified exposure to a particular industry, sector, or group. Rather than purchasing the stocks of a specific group individually, the HOLDRS investor is able to trade one security to gain instant exposure to the desired sector. The investor retains all ownership rights and benefits related to the underlying stocks including voting rights, dividends, and the ability to sell the individual stocks when desired.

HOLDRS offer tax benefits. Unlike other diversified investments, where taxable gains are generated whenever a stock in the portfolio is sold, HOLDRS taxes are not based on someone else's investment decision. HOLDRS, with their normal eight cents per share annual custody charge, have much lower fees than other diversified investments. Additionally, most other diversified investments are priced only once a day while HOLDRS trade throughout the course of the day.

HOLDRS are created when stocks considered representative of a targeted industry, sector, or group are selected on the basis of market capitalization, liquidity, P/E ratio, or other attributes. Once selected for inclusion in the HOLDRS, these stocks may be equally weighted or weighted on a modified market cap basis. Membership and weightings in the HOLDRS portfolio is established prior to the HOLDRS initial public offering. Underlying stocks do not change except due to corporate events such as mergers and acquisitions or spin-offs. When, for example, a spin-off does occur with a member stock, investors will receive that security in their brokerage account outside of the HOLDRS investment.

HOLDRS can be cancelled by instructing the broker to deliver the HOLDRS to the trustee and pay a cancellation fee of up to $10 per round lot of 100 HOLDRS to the trustee. The underlying shares are then transmitted to the investor's account. Canceling the HOLDRS is not a taxabale event, merely a conversion of ownership from within a HOLDR to outside a HOLDR. Once canceled, the component stocks can be sold or held as desired.

BIOTECH HOLDRS SYMBOL (BBH)

Description

The Biotech HOLDRS Trust will issue depositary receipts called biotech HOLDRS SM representing an undivided beneficial ownership in the common stock of a group of specified companies that are involved in various segments of the biotechnology industry. The Bank of New York is the trustee. Biotech HOLDRS may be acquired, held, or transferred in a round-lot amount of 100 biotech HOLDRS or round-lot multiples. Biotech HOLDRS are separate from the underlying deposited common stocks that are represented by the biotech HOLDRS. The Biotech HOLDRS Trust is not a registered investment company under the Investment Company Act of 1940.

Specific share amounts for each round-lot of 100 biotech HOLDRS, as of 6/28/01, are set forth in the table below. The share amounts specified will not change except for changes due to corporate actions.

Name & Symbol	# of Shares	Name & Symbol	# of Shares
Amgen Inc. (AMGN)	46	IDEC Pharma Corp. (IDPH)	12
Genentech, Inc. (DNA)	44	QLT Inc. (QLTI)	5
Biogen, Inc. (BGEN)	13	Millennium Pharma. (MLNM)	12
Immunex Corporation (IMNX)	42	Shire Pharm Grp plc. (SHPGY)	6.8271
Applera Corp- Group (ABI)	18	Affymetrix, Inc. (AFFX)	4
MedImmune, Inc. (MEDI)	15	Human Genome Sciences (GSI)	8
Chiron Corporation (CHIR)	16	ICOS Corporation (ICOS)	4
Genzyme Corporation (GENZ)	14	Enzon, Inc. (ENZN)	3
Gilead Sciences, Inc. (GILD)	8	Applera Corp-Celera Gen (CRA)	4
Sepracor Inc. (SEPR)	6	Alkermes, Inc. (ALKS)	4

BROADBAND HOLDRS SYMBOL (BDH)

Description

The Broadband HOLDRS Trust will issue depositary receipts called broadband HOLDRS representing an undivided beneficial ownership in the common stock of a group of specified companies that, among other things, develop, manufacture, and market products and services that facilitate the transmission of data, video, and voice more quickly and more efficiently than traditional telephone line communications. The Bank of New York is the trustee. Broadband HOLDRS may be acquired, held, or transferred in a round-lot amount of 100 broadband HOLDRS or round-lot multiples. Broadband HOLDRS are separate from the underlying deposited common stocks that are represented by the broadband HOLDRS. The Broadband HOLDRS Trust is not a registered investment company under the Investment Company Act of 1940.

The specific share amounts for each round-lot of 100 broadband HOLDRS, as of 6/28/01, are set forth in the table below. The share amounts specified will not change except for changes due to corporate actions or reconstitution events.

Name & Symbol	# of Shares	Name & Symbol	# of Shares
Lucent Technologies, Inc. (LU)	29	Conexant Systems (CNXT)	2
Nortel Networks (NT)	28	Scientific-Atlanta, Inc. (SFA)	2
JDS Uniphase Corp (JDSU)	11.8	Applied Micro Circuits (AMCC)	2
QUALCOMM, Inc. (QCOM)	8	CIENA Corporation (CIEN)	2
Motorola, Inc. (MOT)	18	Copper Mtn Networks (CMTN)	1
Tellabs, Inc. (TLAB)	4	Next Level Com. (NXTV)	1
Corning, Inc. (GLW)	9	PMC-Sierra, Inc. (PMCS)	1
Sycamore Networks (SCMR)	3	RF Micro Devices. (RFMD)	2
Broadcom Corporation (BRCM)	2	Terayon Com Sys (TERN)	2
Comverse Technology (CMVT)	2		

B2B Internet HOLDRS Symbol (BHH)

Description

The B2B Internet HOLDRS Trust will issue depositary receipts called B2B Internet HOLDRS representing an undivided beneficial ownership in the common stock of a group of specified business to business, or B2B, Internet companies whose products and services are developed for and marketed to other companies who conduct business and electronic commerce on the Internet. The Bank of New York is the trustee. B2B Internet HOLDRS may be acquired, held, or transferred in a round-lot amount of 100 B2B Internet HOLDRS or round-lot multiples. B2B Internet HOLDRS are separate from the underlying deposited common stocks that are represented by the B2B Internet HOLDRS. The B2B Internet HOLDRS Trust is not a registered investment company under the Investment Company Act of 1940.

The specific share amounts for each round-lot of 100 B2B Internet HOLDRS as of 7/16/01, are set forth in the table below. The share amounts specified will not change except for changes due to corporate actions or reconstitution events.

Name & Symbol	# of Shares	Name & Symbol	# of Shares
Internet Capital Group (ICGE)	15	CheckFree Corp (CKFR)	4
Ariba, Inc. (ARBA)	14	Retek Inc. (RETK)	3
Commerce One, Inc. (CMRC)	12	FreeMarkets, Inc. (FMKT)	3
PurchasePro.com, Inc. (PPRO)	4	VerticalNet, Inc. (VERT)	6
SciQuest.com, Inc.(SQST)	3	Scient Corporation (SCNT)	5
QRS Corporation (QRSI)	1	Ventro Corporation (VNTR)	2
Pegasus Solutions (PEGS)	2	Agile Software Corp (AGIL)	4

Europe 2001 HOLDRS Symbol (EKH)

Description

The Europe 2001 HOLDRS Trust will issue depositary receipts called Europe 2001 HOLDRS representing an undivided beneficial ownership in the common stock of a group of specified companies that are among the largest European companies whose equity securities are listed for trading on the New York Stock Exchange or the American Stock Exchange or quoted on the Nasdaq National Market. The Bank of New York is the trustee. Europe 2001 HOLDRS may be acquired held, or transferred in a round-lot amount of 100 Europe 2001 HOLDRS or round-lot multiples. Europe 2001 HOLDRS are separate from the underlying deposited common stocks that are represented by the Europe 2001 HOLDRS. The Europe 2001 HOLDRS Trust is not a registered investment company under the Investment Company Act of 1940.

The specific share amounts for each round-lot of 100 Europe 2001 HOLDRS along with specific share amounts for each round lot of 100 Europe 2001 HOLDRS, as of 7/16/01, are set forth in the table below. The share amounts specified will not change except for changes due to corporate actions or reconstitution events.

Name & Symbol	# of Shares	Name & Symbol	# of Shares
AEGON N.V. (AEGG)	5	KPNQuest N.V. ADR (KQIP)	8
Alcatel - ADS (ALA)	3	Millicom Int'l Celluar S.A. (MICC)	6
Amdocs Limited (DOX)	3	Nokia Corp. ADR (NOK)	5
ARM Holdings Ltd. ADR (ARMHY)	8	Novartis AG ADR (NVS)	5
ASM International NV (ASMI)	13	QIAGEN N.V. (QGENF)	6
ASM Lithography Hldg (ASML)	7	Repsol YPF, S.A.ADR (REP)	11
AstraZeneca P.L.C. ADR (AZN)	4	Royal Dutch Petro Co. (RD)	3
Autonomy Corp PLC ADR (AUTN)	6	Ryanair Holdings (RYAAY)	4
Aventis S.A. ADR (AVE)	2	SAP AG ADR (SAP)	4
AXA Financial, Inc. ADR (AXA)	6	Scottish Power Group (SPI)	7
Bookham Tech PLC ADR (BKHM)	12	Serono S.A. ADR (SRA)	9
BP Amoco P.L.C. ADR (BP)	4	Shire Pharm Grp (SHPGY)	4
Business Objects SA ADS (BOBJ)	4.5	Smartforce Public Ltd. (SMTF)	6
Cable & Wireless PLC ADR (CWP)	4	Sonera Group p.l.c.(SNRA)	9
DaimlerChrysler AG (DCX)	4	STMicroelectronics NV (STM)	4
Deutsche Telekom AG ADR (DT)	5	Telefonica S.A. ADR (TEF)	3.06
Diageo P.L.C. ADR (DEO)	5	Terra Networks S.A (TRLY)	15
Elan Corporation p.l.c. ADR (ELN)	4	Total Fina Elf S.A. ADR (TOT)	3
Ericsson, L.M. Tel ADR (ERICY)	16	UBS AG (UBS)	3
GlaxoSmithKline p.l.c.ADR (GSK)	6	Unilever N.V. NY Shares (UN)	3
Infineon Tech AG ADR (IFX)	5	United Pan-Europe Com (UPCOY)	13
ING Group NV ADR (ING)	4	Vivendi Univl S.A. ADR (V)	3
IONA Tech p.l.c. ADR (IONA)	3	Vodafone Airtouch PLC (VOD)	6
Jazztel p.l.c. ADR (JAZZ)	11	WPP Group plc ADR (WPPGY)	3
Koninklijke Phlps Elc (PHG)	5		

Internet HOLDRS Symbol (HHH)

Description

The Internet HOLDRS Trust will issue depositary receipts called Internet HOLDRS representing an undivided beneficial ownership in the common stock of a group of specified companies that are involved in various segments of the Internet industry. The Bank of New York is the trustee. Internet HOLDRS may be acquired, held, or transferred in a round-lot amount of 100 Internet HOLDRS or round-lot multiples. Internet HOLDRS are separate from the underlying deposited common stocks that are represented by the Internet HOLDRS. The Internet HOLDRS Trust is not a registered investment company under the Investment Company Act of 1940.

The specific share amounts for each round-lot of 100 Internet HOLDRS as of 6/28/01, are set forth in the table below. The share amounts specified will not change except for changes due to corporate actions or reconstitution events.

Name & Symbol	# of Shares	Name & Symbol	# of Shares
AOL Time Warner Inc. (AOL)	42	RealNetworks, Inc. (RNWK)	8
Yahoo Inc. (YHOO)	26	Exodus Comm Inc. (EXDS)	16
Amazon.com Inc. (AMZN)	18	E*Trade Group Inc. (ET)	12
eBay Inc. (EBAY)	12	DoubleClick Inc. (DCLK)	4
At Home Corp. (ATHM)	17	Ameritrade Hldg Corp.(AMTD)	9
Priceline.Com Inc. (PCLN)	7	CNET Networks, Inc. (CNET)	4
CMGI Inc. (CMGI)	10	Network Associates, Inc.(NETA)	7

Internet Architecture HOLDRS Symbol (IAH)

Description

The Internet Architecture HOLDRS Trust will issue depositary receipts called Internet architecture HOLDRS representing an undivided beneficial ownership in the common stock of a group of specified companies that, among other things, develop and market hardware and software designed to enhance the speed and efficiency of connections within and to the Internet, connections within a company's internal networks, and end user access to networks. The Bank of New York is the trustee. Internet architecture HOLDRS may be acquired, held, or transferred in a round-lot amount of 100 Internet Architecture HOLDRS or round-lot multiples. Internet Architecture HOLDRS are separate from the underlying deposited common stocks that are represented by the Internet Architecture HOLDRS. The Internet Architecture HOLDRS Trust is not a registered investment company under the Investment Company Act of 1940.

The specific share amounts for each round-lot of 100 Internet Architecture HOLDRS as of 6/28/01, are set forth in the table below. The share amounts specified will not change except for changes due to corporate actions or reconstitution events.

Name & Symbol	# of Shares	Name & Symbol	# of Shares
Cisco Systems, Inc. (CSCO)	26	Foundry Networks (FDRY)	1
International Busi Mach Corp. (IBM)	13	Gateway, Inc. (GTW)	2
Hewlett-Packard Co. (HWP)	14	Network Applia (NTAP)	2
Sun Microsystems, Inc. (SUNW)	25	Apple Computer (AAPL)	2
EMC Corp. (EMC)	16	Roxio, Inc (ROXI)	0.1646
Dell Computer Corporation (DELL)	19	Veritas Software (VRTS)	0.893
Compaq Computer Corp (CPQ)	13	Extreme Networks (EXTR)	2
Sycamore Networks, Inc. (SCMR)	2	CIENA Corporation (CIEN)	2
Juniper Networks, Inc. (JNPR)	2	Unisys Corp. (UIS)	2
3Com Corporation (COMS)	3	Adaptec, Inc. (ADPT)	1
McDATA Corporation (MCDTA)	0.58891		

Internet Infrastructure HOLDRS Symbol (IIH)

Description

The Internet Infrastructure HOLDRS Trust will issue depositary receipts called Internet infrastructure HOLDRS representing an undivided beneficial ownership in the common stock of a group of specified companies that, among other things, provide software and technology services that enhance Internet content and functionality, network performance, and web site service and analysis to Internet companies. The Bank of New York is the trustee. Internet Infrastructure HOLDRS may be acquired, held or transferred in a round-lot amount of 100 Internet infrastructure HOLDRS or round-lot multiples. Internet Infrastructure HOLDRS are separate from the underlying deposited common stocks that are represented by the Internet Infrastructure HOLDRS. The Internet Infrastructure HOLDRS Trust is not a registered investment company under the Investment Company Act of 1940.

The specific share amounts for each round-lot of 100 Internet Infrastructure HOLDRS as of 6/28/01, are set forth in the table below. The share amounts specified will not change except for changes due to corporate actions or reconstitution events.

Name & Symbol	# of Shares	Name & Symbol	# of Shares
Exodus Communications (EXDS)	12	Portal Software, Inc. (PRSF)	6
Akamai Technologies (AKAM)	3	Vitria Technology, Inc. (VITR)	4
VeriSign, Inc. (VRSN)	6.15	InterNAP Network Svcs (INAP)	5
InfoSpace.com Inc. (INSP)	8	Digital Island, Inc. (ISLD)	2
BroadVision, Inc. (BVSN)	9	Kana Communications (KANA)	2
Vignette Corporation (VIGN)	6	Usinternetworking, Inc. (USIX)	3
Inktomi Corp. (INKT)	4	E.piphany, Inc. (EPNY)	1.5
BEA Systems, Inc. (BEAS)	10	NaviSite, Inc. (NAVI)	2
RealNetworks, Inc. (RNWK)	6	Openwave Systems (OPWV)	3.221

Appendix A — Fund Pages

Market 2000 + HOLDRS Symbol (MKH)

Description

The Market 2000+ HOLDRS Trust will issue depositary receipts called Market 2000+ HOLDRS representing an undivided beneficial ownership in the common stock of a group of 51 specified companies that are among the largest companies whose commons stock or American depository shares are listed for trading on the New York Stock Exchange or The American Stock Exchange or quoted on the Nasdaq. The Bank of New York is the trustee. Market 2000+ HOLDRS may be acquired, held, or transferred in a round-lot amount of 100 Market 2000+ HOLDRS or round-lot multiples. Market 2000+ HOLDRS are separate from the underlying deposited common stocks that are represented by the Market 2000+ HOLDRS. The Market 2000+ HOLDRS Trust is not a registered investment company under the Investment Company Act of 1940.

The specific share amounts for each round-lot of 100 Market 2000+ HOLDRS as of 7/16/01, are set forth in the table below. The share amounts specified will not change except for changes due to corporate actions or reconstitution events.

Name & Symbol	# of Shares	Name & Symbol	# of Shares
America Online, Inc. (AOL)	6	LM Ericsson Tele Co. (ERICY)	9
American Int'l Group, (AIG)	2	Lucent Technologies (LU)	4
Astrazeneca p.l.c. (AZN)	4	McDATA Corp (MCDTA)	0.073614
AT & T Corp. (T)	6	Merck & Co., Inc. (MRK)	3
AT & T Wireless Svcs. (AWE)	1.9308	Microsoft (MSFT)	3
Avaya, Inc. (AV)	0.3333	Morgan Stanley DW (MWD)	2
BellSouth Corporation (BLS)	5	NTT Data Corporation (NTT)	3
BP Amoco p.l.c. (BP)	3	Nokia Corp. (NOK)	4
Bristol-Meyers Squibb (BMY)	3	Nortel Networks (NT)	2
British Telecom (BTY)	2	Novartis AG (NVS)	5
Cisco Systems, Inc. (CSCO)	3	Oracle Corporation (ORCL)	4
Citigroup (C)	3	Pfizer Inc. (PFE)	4
The Coca-Cola Company (KO)	3	Qwest Communications Int'l (Q)	4
Dell Computer Corp (DELL)	5	Royal Dutch Petroleum Co. (RD)	3
Deutsche Telekom AG (DT)	5	SBC Communications Inc. (SBC)	4
Eli Lilly and Company (LLY)	2	Sony Corporation (SNE)	2
EMC Corporation (EMC)	2	Sun Microsystems, Inc. (SUNW)	4
Exxon Mobil Corp (XOM)	2	Syngenta (SYT)	1.038609
France Telecom (FTE)	2	Texas Instruments Inc (TXN)	3
General Electric Co. (GE)	3	Total Fina Elf S.A. (TOT)	2
Glaxo Smith Kline P.l.c (GSK)	3	Toyota Motor Corporation (TM)	2
Hewlett-Packard Co. (HWP)	4	Verizon Communications (VZ)	4
Home Depot, Inc. (HD)	4	Viacom Inc. (VIA.B)	3
Intel Corporation (INTC)	2	Vodafone Airtouch p.l.c. (VOD)	5
International Bus Mach (IBM)	2	Wal-Mart Stores Inc. (WMT)	4
JDS Uniphase (JDSU)	2	WorldCom, Inc. (WCOM)	5
Johnson & Johnson (JNJ)	4	WorldCom. - MCI Group (MCIT)	0.2

OIL SERVICE HOLDRS SYMBOL (OIH)

Description

The Oil Service HOLDRS Trust will issue depositary receipts called oil service HOLDRS representing an undivided beneficial ownership in the common stock of a group of specified companies that, among other things, provide drilling, well site management, and related products and services for the oil service industry. The Bank of New York is the trustee. Oil Service HOLDRS may be acquired, held, or transferred in a round-lot amount of 100 Oil Service HOLDRS or round-lot multiples. Oil Service HOLDRS are separate from the underlying deposited common stocks that are represented by the Oil Service HOLDRS. The Oil Service HOLDRS Trust is not a registered investment company under the Investment Company Act of 1940.

The specific share amounts for each round-lot of 100 oil service HOLDRS as of 6/28/01, are set forth in the table below. The share amounts specified will not change except for changes due to corporate actions or reconstitution events.

Name & Symbol	# of Shares	Name & Symbol	# of Shares
Baker Hughes Inc. (BHI)	21	Noble Drilling Corp. (NE)	11
BBJ Services Co. (BJS)	14	National-Oil Well Inc. (NOI)	7
Cooper Cameron Corp. (CAM)	4	Rowan Cos Inc. (RDC)	8
Diamond Offshore Drilling (DO)	11	Transocean Sed Forex (RIG)	18
Ensco International Inc. (ESV)	11	Santa Fe Int'l Corp. (SDC)	10
Global Marine Inc. (GLM)	15	Smith International Inc. (SII)	4
Grant Prideco Inc. (GRP)	9	Schlumberger Ltd. (SLB)	11
Halliburton Co. (HAL)	22	Tidewater Inc. (TDW)	5
Hanover Compressor Co. (HC)	5	Weatherford Intl. Inc. (WFT)	9
Nabors Industries (NBR)	12		

PHARMACEUTICAL HOLDRS SYMBOL (PPH)

Description

The Pharmaceutical HOLDRS Trust will issue depositary receipts called pharmaceutical HOLDRS representing an undivided beneficial ownership in the common stock of a group of specified companies that are involved in various segments of the pharmaceutical industry. The Bank of New York is the trustee. Pharmaceutical HOLDRS may be acquired, held, or transferred in a round-lot amount of 100 pharmaceutical HOLDRS or round-lot multiples. Pharmaceutical HOLDRS are separate from the underlying deposited common stocks that are represented by the pharmaceutical HOLDRS. The Pharmaceutical HOLDRS Trust is not a registered investment company under the Investment Company Act of 1940.

The specific share amounts for each round-lot of 100 pharmaceutical HOLDRS as of 6/28/01, are set forth in the table below. The share amounts specified will not change except for changes due to corporate actions or reconstitution events.

Name & Symbol	# of Shares	Name & Symbol	# of Shares
Merck & Co., Inc. (MRK)	22	King Pharmaceuticals (KG)	3.187
Pfizer Inc. (PFE)	58	Forest Laboratories, Inc. (FRX)	2
Johnson & Johnson (JNJ)	25	Andrx Corporation (ADRX)	2
Bristol-Meyers Squibb (BMY)	18	Allergan, Inc. (AGN)	1
Eli Lilly & Company (LLY)	10	Watson Pharmaceuticals (WPI)	1
Schering-Plough Corp (SGP)	14	ICN Pharmaceuticals (ICN)	1
American Home Prod. (AHP)	12	Mylan Laboratories, Inc. (MYL)	1
Abbot Laboratories (ABT)	14	IVAX Corporation (IVX)	1.875
Biovail Corporation Int'l (BVF)	4		

REGIONAL BANK HOLDRS SYMBOL (RKH)

Description

The Regional Bank HOLDRS Trust will issue depositary receipts called regional bank HOLDRS representing an undivided beneficial ownership in the common stock of a group of specified companies that are involved in various segments of the regional banking industry. The Bank of New York is the trustee. Regional Bank HOLDRS may be acquired, held or transferred in a round-lot amount of 100 Regional Bank HOLDRS or round-lot multiples. Regional Bank HOLDRS are separate from the underlying deposited common stocks that are represented by the Regional Bank HOLDRS. The Regional Bank HOLDRS Trust is not a registered investment company under the Investment Company Act of 1940.

The specific share amounts for each round-lot of 100 regional bank HOLDRS as of 6/28/01, are set forth in the table below. The share amounts specified will not change except for changes due to corporate actions or reconstitution events.

Name & Symbol	# of Shares	Name & Symbol	# of Shares
AmSouth Bancorp (ASO)	12	Northern Trust Corp (NTRS)	7
BB&T Corp. (BBT)	10	Bank One Corp. (ONE)	33
Comercia Incorporated (CMA)	5	PNC Financial Svcs Grp (PNC)	9
FleetBoston Financial (FBF)	25	Synovus Financial Corp. (SNV)	8
Fifth Third Bancorp (FITB)	13.5	SunTrust Banks Inc. (STI)	9
First Union Corporation (FTU)	29	State Street Corp. (STT)	10
KeyCorp (KEY)	13	U.S. Bancorp (USB)	56.83
Mellon Financial Corp (MEL)	14	Wachovia Corporation (WB)	6
Marshall & Ilsley Corp (MI)	3	Wells Fargo & Co. (WFC)	24

Appendix A — Fund Pages

Retail HOLDRS Symbol (RTH)

Description

The Retail HOLDRS Trust will issue depositary receipts called retail HOLDRS representing an undivided beneficial ownership in the common stock of a group of specified companies that are involved in the retailing industry. The Bank of New York is the trustee. Retail HOLDRS may be acquired, held or transferred in a round-lot amount of 100 retail HOLDRS or round-lot multiples. Retail HOLDRS are separate from the underlying deposited common stocks that are represented by the retail HOLDRS. The Retail HOLDRS Trust is not a registered investment company under the Investment Company Act of 1940.

The specific share amounts for each round-lot of 100 retail HOLDRS as of 7/16/01, are set forth in the table below. The share amounts specified will not change except for changes due to corporate actions or reconstitution events.

Name & Symbol	# of Shares	Name & Symbol	# of Shares
Albertsons Inc. (ABS)	8	Lowe's Companies (LOW)	14
Amazon.Com Inc. (AMZN)	7	Limited Inc. (LTD)	8
Best Buy Co.Inc. (BBY)	4	May Department Stores (MAY)	6
Costco Wholesale (COST)	8	Radioshack Corp. (RSH)	3
CVS Corp. (CVS)	7	Sears Roebuck & Co. (S)	6
Federated Department (FD)	4	Safeway Inc. (SWY)	9
GAP Inc. (GPS)	16	Target Corp. (TGT)	16
Home Depot Inc. (HD)	40	TJX Companies Inc. (TJX)	5
Kroger Co. (KR)	15	Walgreen Co. (WAG)	19

Semiconductor HOLDRS Symbol (SMH)

Description

The Semiconductor HOLDRS Trust will issue depositary receipts called semiconductor HOLDRS representing an undivided beneficial ownership in the common stock of a group of specified companies that, among other things, develop, manufacture, and market integrated circuitry and other products made from Semiconductors which allow for increased speed and functionality in components for computers and other electronic devices. The Bank of New York is the trustee. Semiconductor HOLDRS may be acquired, held, or transferred in a round-lot amount of 100 Semiconductor HOLDRS or round-lot multiples. Semiconductor HOLDRS are separate from the underlying deposited common stocks that are represented by the Semiconductor HOLDRS. The Semiconductor HOLDRS Trust is not a registered investment company under the Investment Company Act of 1940.

The specific share amounts for each round-lot of 100 Semiconductor HOLDRS are set forth in the table below and were determined on 12/15/99 so that the initial weightings of each stock approximated the relative market capitalizations of the specified companies, subject to a maximum weight of 20 percent. The share amounts specified will not change except for changes due to corporate actions or reconstitution events. Components listed as of 6/28/01.

Name & Symbol	# of Shares	Name & Symbol	# of Shares
Intel Corporation (INTC)	30	Altera Corporation (ALTR)	6
Texas Instruments Inc. (TXN)	22	Linear Technology Corp (LLTC)	5
Applied Materials, Inc. (AMAT)	13	Vitesse Semiconductor (VTSS)	3
Broadcom Corp Cl.A (BRCM)	2	Advanced Micro Device (AMD)	4
Micron Technology, Inc. (MU)	9	KLA-Tencor Corp. (KLAC)	3
Analog Devices Inc. (ADI)	6	National Semiconductor (NSM)	3
Xilinx, Inc. (XLNX)	5	Atmel Corporation (ATML)	8
Maxim Integrated Pdts. (MXIM)	5	Novellus Systems Inc. (NVLS)	2
LSI Logic Corporation (LSI)	5	Amkor Tech, Inc. (AMKR)	2

Appendix A — Fund Pages

Software **HOLDRS** Symbol **(SWH)**

Description

The Software HOLDRS Trust will issue depositary receipts called Software HOLDRS representing an undivided beneficial ownership in the common stock of a group of specified companies that are involved in various segments of the software industry. The Bank of New York is the trustee. Software HOLDRS may be acquired held or transferred in a round-lot amount of 100 Software HOLDRS or round-lot multiples. Software HOLDRS are separate from the underlying deposited common stocks that are represented by Software HOLDRS. Software HOLDRS Trust is not a registered investment company under the Investment Company Act of 1940.

The specific share amounts for each round-lot of 100 Software HOLDRS are set forth in the table below. The share amounts specified will not change except for changes due to corporate actions or reconstitution events. Components listed as of 6/28/01.

Name & Symbol	# of Shares	Name & Symbol	# of Shares
Adobe Systems Inc. (ADBE)	6	Nuance Communi (NUAN)	1
BMC Software Inc. (BMC)	7	Openwave Systems (OPWV)	2
Computer Associates Intl. (CA)	17	Oracle Corp. (ORCL)	24
Check Point Software (CHKP)	6	PeopleSoft Inc. (PSFT)	8
Intuit Inc. (INTU)	6	Rational Software Corp. (RATL)	5
I2 Technologies Inc. (ITWO)	10	SAP AG-ADR (SAP)	16
Macromedia Inc. (MACR)	1	Sapient Corp. (SAPE)	3
Mercury Interactive (MERQ)	2	Siebel Systems Inc. (SEBL)	8
Microsoft Corp. (MSFT)	15	Tibco Software Inc. (TIBX)	5
Micromuse Inc. (MUSE)	2	Veritas Software Co. (VRTS)	7

Telecom HOLDRS Symbol (TTH)

Description

The Telecom HOLDRS Trust will issue depositary receipts called Telecom HOLDRS representing an undivided beneficial ownership in the common stock of a group of 20 specified companies that are involved in various segments of the telecommunications industry. The Bank of New York is the trustee. Telecom HOLDRS may be acquired, held, or transferred in a round-lot amount of 100 Telecom HOLDRS or round-lot multiples. Telecom HOLDRS are separate from the underlying deposited common stocks that are represented by the Telecom HOLDRS. The Telecom HOLDRS Trust is not a registered investment company under the Investment Company Act of 1940.

The specific share amounts for each round-lot of 100 Telecom HOLDRS are set forth in the table below and were determined on 12/15/99 so that the initial weightings of each stock approximated the relative market capitalizations of the specified companies, subject to a maximum weight of 20 percent. The share amounts specified will not change except for changes due to corporate actions or reconstitution events. Component list as of 7/16/01.

Name & Symbol	# of Shares	Name & Symbol	# of Shares
SBC Communications (SBC)	27	Nextel Communications (NXTL)	6
AT & T Corp. (T)	25	WorldCom Group (WCOM)	22
Qwest Commun Int'l. (Q)	12.9173	Level 3 Communication (LVLT)	3
BellSouth Corp. (BLS)	15	ALLTEL Corp. (AT)	2
Telephone & Data Sys, (TDS)	1	BCE Inc. (BCE)	5
NTL Inc. (NTI)	1.25	Sprint Corp. (FON Grp) (FON)	6
Broadwing Inc. (BRW)	2	Sprint Corp. (PCS Grp) (PCS)	6
Global Crossing Ltd. (GX)	6	McLeodUSA Inc. (MCLD)	3
Verizon Communications (VZ)	21.76	Century Tele Enterprises(CTL)	1

Utilities HOLDRS Symbol (UTH)

Description

The Utilities HOLDRS Trust will issue depositary receipts called utilities HOLDRS representing an undivided beneficial ownership in the common stock of a group of specified companies that are involved in various segments of the utilities industry. The Bank of New York is the trustee. Utilities HOLDRS may be acquired, held, or transferred in a round-lot amount of 100 Utilities HOLDRS or round-lot multiples. Utilities HOLDRS are separate from the underlying deposited common stocks that are represented by the Utilities HOLDRS. The Utilities HOLDRS Trust is not a registered investment company under the Investment Company Act of 1940.

The specific share amounts for each round-lot of 100 Utilities HOLDRS are set forth in the table below. The share amounts specified will not change except for changes due to corporate actions or reconstitution events. Components are listed as of 6/28/01.

Name & Symbol	# of Shares	Name & Symbol	# of Shares
American Electric Power (AEP)	14	Exelon Corporation (EXC)	15
Progress Energy Inc. (PGN)	7	FirstEnergy Corp. (FE)	10
Dominion Resources Inc. (D)	11	FPL Group, Inc. (FPL)	8
Duke Energy Corp (DUK)	30	Mirant Corporation (MIR)	11.5
Dynergy Inc. (DYN)	12	PG&E Corp. (PCG)	17
Consolidated Edison (ED)	9	Public Svc Enterpr Grp (PEG)	10
Edison International (EIX)	15	Reliant Energy (REI)	13
Enron Corp. (ENE)	12	Southern Company (SO)	29
El Paso Energy Corp (EPG)	10	TXU Corp. (TXU)	12

Wireless HOLDRS Symbol (WWH)

Description

The Wireless HOLDRS℠ Trust will issue depositary receipts called Wireless HOLDRS℠ representing an undivided beneficial ownership in the common stock of a group of specified companies that are involved in various segments of the wireless industry. The Bank of New York is the trustee. Wireless HOLDRS may be acquired, held, or transferred in a round-lot amount of 100 Wireless HOLDRS or round-lot multiples. Wireless HOLDRS are separate from the underlying deposited common stocks that are represented by the Wireless HOLDRS. The Wireless HOLDRS Trust is not a registered investment company under the Investment Company Act of 1940.

The specific share amounts for each round-lot of 100 Wireless HOLDRS are set forth in the table below. The share amounts specified will not change except for changes due to corporate actions or reconstitution events. Components are listed as of 6/28/01.

Name & Symbol	# of Shares	Name & Symbol	# of Shares
LM Ericsson Tele(ERICY)	74	Deutsche Telecom AG (DT)	18.5
AT&T Wireless (AWE)	40	RF Micro Devices, Inc. (RFMD)	4
Motorola, Inc. (MOT)	41	Crown Castle Int'l (CCI)	4
Nokia Corp. (NOK)	23	Nextel Partners, Inc. (NXTP)	4
Vodafone Group p.l.c. (VOD)	21	Telesp Celular Partic SA (TCP)	3
Sprint Corp-PCS Group (PCS)	21	Western Wireless (WWCA)	2
Verizon Communications (VZ)	17	Research in Motion Ltd (RIMM)	2
SK Telecom Co., Ltd. (SKM)	17	US Cellular Corp (USM)	1
Nextel Commun (NXTL)	16	Aether Systems, Inc. (AETH)	1

Appendix B — The SuperTrust

THE SUPERTRUST TRUST FOR CAPITAL MARKET FUND INC. SHARES

As described in Chapter 1, the small investment management firm of Leland O'Brien Rubinstein Associates Incorporated and its subsidiary, SuperShare Services Corporation, began creating a new financial product immediately following the stock market "Crash" of October 1987.

On November 5, 1992, the "SuperTrust" product was finally launched, following five years of development and marketing. A billion dollars of assets invested by institutional investors and more than 400 individuals constituted the initial subscriptions.

The purposes of the SuperTrust included:

- The ability to trade the S&P 500 Index as a single security on an exchange
- The ability to separate index units (both S&P 500 and Money Market) into component shares
- The ability to hold various combinations of the units and component shares in order to achieve a variety of investment return payoffs over a three-year time horizon

The investment return payoffs would be fully backed by assets of investment companies registered under the Investment Company Act of 1940.

The following sections briefly describe the structure, the securities, the payoffs, and the actual outcomes of the SuperTrust product.

STRUCTURE

The base level for the product was an open-end mutual fund, Capital Market Fund, Inc. (the Fund) that had two series of shares. The Index Series was composed of an S&P 500 Index fund, managed initially by Wells Fargo Nikko Investment Advisers and later by Bank of New York. The U.S. Treasury Money Market Series consisted of a U.S. Treasury money market fund, managed by Western Asset Management Company.

The next level in the structure was a unit investment trust, formally The SuperTrust Trust for Capital Market Fund Inc. Shares (the SuperTrust). It had two subtrusts, the Index Trust and the Money Market Trust. Shares of the Index Series and the Money Market Series of the Fund could be deposited into the respective subtrusts in exchange for redeemable units of beneficial interest in such subtrusts.

Securities

The SuperTrust product consisted of six exchange-traded securities, as follows:

- The Index Trust SuperUnit, issued by the Index Trust in exchange for the deposit of Index Shares of the Index Series of the Fund
- The Money Market SuperUnit, issued by the Money Market Trust in exchange for the deposit of Money Market Shares of the Money Market Series of the Fund
- The Index SuperUnit could be separated into two component securities:
  - The Priority SuperShare
  - The Appreciation SuperShare
- The Money Market SuperUnit could be separated into two component securities:
  - The Income and Residual SuperShare
  - The Protection SuperShare

The two SuperUnits traded on the American Stock Exchange. The four SuperShares traded on the Chicago Board Options Exchange.

Payoffs

Three years after launch, the SuperTrust would (and did) terminate. The payoffs to the six securities were defined by the following relationships:

- The Index SuperUnit received all income during the life of the Index Trust, and the net asset value (NAV) of the Index SuperUnit at termination. The initial NAV for the Index SuperUnit was $100.
- The Money Market SuperUnit received all income during the life of the Money Market Trust, and the NAV of the Money Market SuperUnit at termination. The initial NAV for the Money Market SuperUnit was $50.

The payoffs to each of the SuperShares was determined by the **NAV of the Index SuperUnit at termination,** as follows:

- The Priority SuperShare received all income from the Index SuperUnit during the life of the Index Trust and the termination NAV of the Index SuperUnit **up to $125.**
- The Appreciation SuperShare received the termination NAV of the Index SuperUnit **in excess of $125.**
- The Protection SuperShare would receive an amount equal to the extent to which the NAV of the Index SuperUnit at termination **was less than $100.** The maximum amount the Protection SuperShare could be paid was $30.
- The Income and Residual SuperShare would receive all income from the Money Market SuperUnit during the life of the Money Market Trust and an amount equal to $50, *less* the amount paid to the Protection SuperShare at termination.

Appendix B — The SuperTrust

The structure, the securities, the initial pricing and payoffs are summarized graphically below.

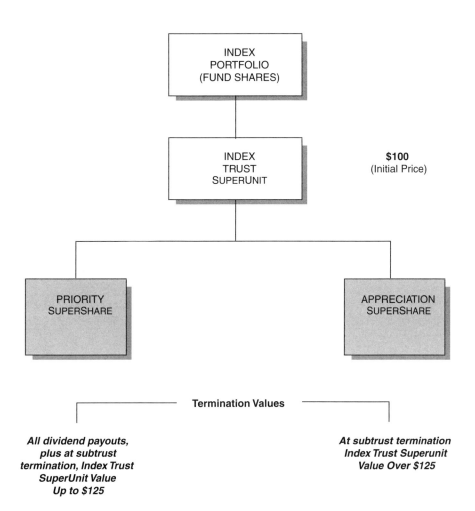

Power Investing with Basket Securities

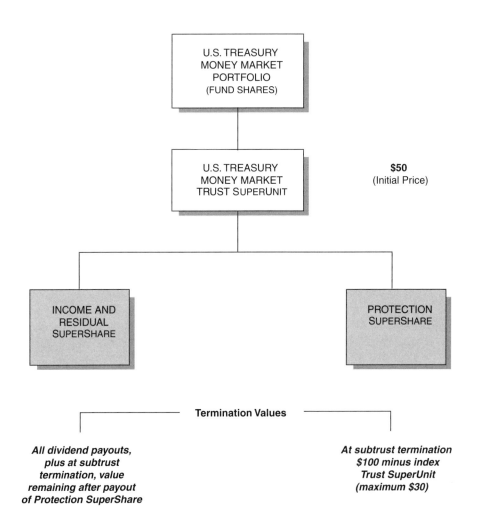

Initial Pricing & Termination Values

Actual Outcomes

On November 5, 1995, at the termination of the SuperTrust following three years of operations, the NAV of the Index SuperUnit was approximately $140, and the NAV of the Money Market SuperUnit was $50. Accordingly, on this basis the payouts to each of the six securities were as follows:

- Index SuperUnit $140.00
- Money Market SuperUnit $50.00
- Priority SuperShare $125.00
- Appreciation Supershare $15.00
- Income and Residual SuperShare $50.00
- Protection SuperShare $0.00

Appendix C — The Active/Passive Indicator

This brief primer is for those readers who desire assistance in setting up a point-and-figure based active/passive indicator. It merely supplies the most fundamental explanations but is more than sufficient to begin. There are a number of excellent resources that explain the history, creation, and interpretation of these interesting and highly reliable charts. The two most essential works are, *The Three-Point Reversal Method of Point and Figure*, by Michael L. Burke available from Chartcraft Inc, New Rochelle, NY, and *Point & Figure Charting*, by Thomas J. Dorsey, published by Wiley Finance Editions.

A POINT-AND-FIGURE CHART PRIMER

There are many different kinds of stock charts and all serve the same function — to record the struggle between bulls and bears. Charts concentrate on differing time periods — hours, days, weeks, and months. The principles are similar for each time scale. Classical charting requires only paper and a pencil. While computers speed up the process, they can deprive you of an intuitive "hands-on" feel.

There are two basic chart types: bar charts and point-and-figure charts. The Basket Case Active/Passive Indicator uses a point-and-figure (P&F) relative strength chart to determine the percentage allocation to active or passive portions of the equity portfolio.

POINT-AND-FIGURE CHARTS

Point-and-figure charts provide an excellent way to monitor supply and demand. A P&F chart is an organized way of recording whether buyers outnumber sellers, or vice versa, over time. On a P&F chart an 'X' represents rising demand (more buyers than sellers, pushing prices higher), while an 'O' represents rising supply (more sellers than buyers, forcing prices down). X's and O's are plotted according to the price movement of the security.

P&F charts are the oldest form of stock charting and the only charting method original to the market. They afford clear buy and sell signals for individual securities and can also be used to analyze the market itself.

CONSTRUCTING P&F CHARTS

Constructing P&F charts is easy. Graph paper is used, and a value is assigned to each box. Share price value runs on the vertical, or y, axis while time is on the horizontal, or *x*, axis. Time is not plotted on a rigorous basis, only when a price change occurs. Volume is not recorded at all.

On P&F charts each box is worth a predetermined value established by the chart scale. There are different scales for different investments. The active/passive indicator uses the following scale:

Over $100	$1.00 per box
Between $40 and $100	40¢ per box
Between $10 and $40	20¢ per box
Between $2.50 and $10.00	10¢ per box
Between $1.50 and $2.50	2¢ per box
Under $1.50	1¢ per box

Point-and-figure charts are composed of alternating columns of X's and O's. X's denote rising prices and O's signify falling prices. X's and O's are never in the same column, but in alternate columns. Each column shows prices moving up, or prices moving down. It cannot be both. Each time there is a change in price direction, the chart moves one column to the right. Note the chart in Figure C.1.

Figure C.1 is a one-year chart of the S&P SPDR Index 500 Fund (SPY). The beginning point on the chart, at the extreme left, tells us that the fund sold for $123 per share and that each box is worth $1.00. The second column shows prices falling to $122, indicated with a column of O's. The third column rises in a column of X's to $126 before falling in a column of O's back to $123, forming a rising bottom. Time on a point-and-figure chart is plotted using the numbers 1 through 12 to denote

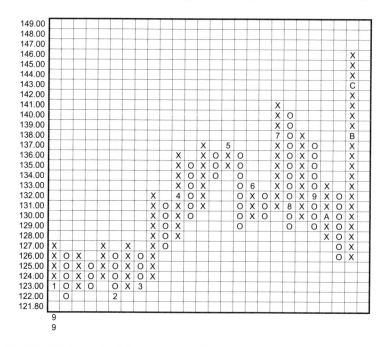

FIGURE C.1 Example Point-and-Figure Chart

months of the year. The letters A-B-C are used for October, November, and December if the spreadsheet is maintained on a computer.

When prices are rising, the current column on the chart will be filled with a series of ascending X's. Each day, check the fund's closing price. If it is higher than the previous day and high enough to fill another box, do so and go on to your next chart. Before plotting another X, the entire value of the next higher box is filled. For example, if the last price was $100 in a column of X's, to move to the next higher box the price must be at least $101. If the day's high was $100.65, it will not be high enough to warrant filling in the next box. This filters out a degree of market noise. As long as the price continues to rise, be sure it moves high enough to fill the next box value.

When the current price is not sufficient to fill in the next higher box, leave the plot alone. But, when the price reverses and falls below the previous day, see if it is sufficient to cause a three-box reversal. For example, in the first column of our chart of the price reach a high of $127. In order to move over to the next column and begin a downward series of O's, the price had to reverse by the value of three boxes, or from $127 to at least $124. If the price had fallen to $124.75, we would ignore the move and do nothing because it was insufficient to trigger a full three-box reversal. Once again, you have filtered out market noise.

When the daily NAV stops falling and closes higher than the previous day, you must determine if the upward price change is sufficient to trigger a three-box reversal to the upside. In the second column from the left in Figure C.1, a three-box reversal takes place off the low of $122. The new price was $125, triggering a three-box upside reversal sufficient to move over one column to the right and begin a column of ascending X's.

Consider what happens when a basket fund's price falls below a scale boundary. If the price is $10.40 and trending up in a column of X's, the price would have to fall to $9.90 to trigger a three-box reversal into a column of O's (see Table C.1).

Or, if the fund is trending down and the last entry is $39.60, a three-box reversal would take the price back up to $40.40 ($39.80-$40.00-$40.40) (see Table C.2).

A three box-reversal at $101 would take the price down to $99 ($102-$100-$99 (see Table C.3).

TABLE C.1
Three-Box Reversal to the Downside: $10.40 to $9.90

10.60				
10.40	X			
10.20	X	O		Price at $10.20
10.00	X	O		Price at $10.00
9.90	X	O		Price at $9.90
9.80	X			
9.70				

TABLE C.2
Three-Box Reversal to the Upside: $39.60 to $40.40

40.80			
40.40	O	X	Price at $40.40
40.00	O	X	Price at $40.00
39.80	O	X	Price at $39.80
39.60	O		Price at $39.60
39.40			

TABLE C.3
Three-Box Reversal to the Downside: $101.00 to $99.20

102.00			
101.00	X		ETF price at $101.00
100.00	X	O	ETF price at $100.00
99.60	X	O	ETF price at $99.60
99.20	X	O	ETF price at $99.20
88.80	X		ETF price at $88.80
88.40			ETF price at $88.40

Figures C.2 and C.3 show falling-price and rising-price flow charts.

ESSENTIAL CHART PATTERNS

Point-and-figure charts reveal the changing tide of investor demand. When demand exceeds supply, prices will rise - denoted with a column of rising X's. When supply exceeds demand, prices fall, appearing on the chart as a column of descending O's.

Chart patterns, because they record market behavior, tend to form into repetitive groups. There are 16 essential point-and-figure chart patterns, 8 buy signals and 8 sell signals (see Figure C.4).

RELATIVE STRENGTH BUY/SELL SIGNALS

Relative strength charts share the same buy and sell signals with price charts, but give these signals considerably less often. Relative strength buy and sell signals last significantly longer than price chart signals. A price chart buy or sell signal may last as little as a day, but relative strength signals can last for up to two or more years. So, even though buy and sell signals are exactly the same for relative strength point-and-figure charts as they are for price charts, they warrant much greater attention. Since the passive/active indicator will be using the buy-sell signals that follow, merely remember that your charts will be much slower to speak, and warrant even more attention.

Appendix C — The Active/Passive Indicator

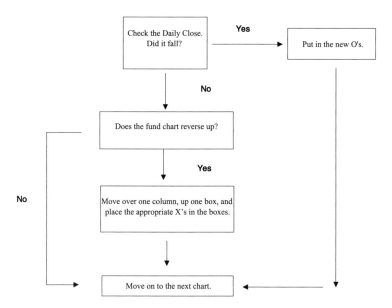

FIGURE C.2 Point-and-Figure Falling Price Flow Chart

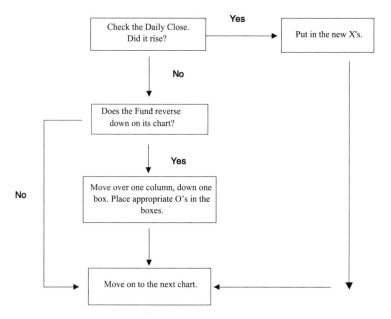

FIGURE C.3 Point-and-Figure Rising Price Flow Chart

Basic Buy Signals

1. Basic Buy Signal

		X	
X		X	
X	O	X	
X	O	X	
	O		

2. Basic Buy Signal with a Rising Bottom

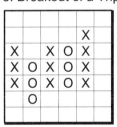

Basic Sell Signals

1. Basic Sell Signal

	X		
O	X	O	
O	X	O	
O		O	

2. Basic Sell Signal with a Declining Top

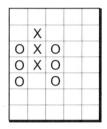

Complex Buy Signals

3. Breakout of a Triple Top

			X	
X		X	O	X
X	O	X	O	X
X	O	X	O	X
	O			

4. Ascending Triple Top

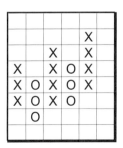

Complex Sell Signals

3. Breakout of a Triple Bottom

	X		X	
O	X	O	X	O
O	X	O	X	O
O		O		O
				O

4. Descending Triple Bottom

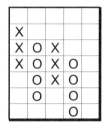

FIGURE C.4 Essential Point-and-Figure Chart Patterns

Appendix C — The Active/Passive Indicator 233

5. Spread Triple Top

						X
X		X				X
X	O	X	O	X		X
X	O	X	O	X	O	X
	O		O	X	O	X
				O		O

5. Spread Triple Bottom

	X			X		
	X		X	O	X	O
O	X	O	X	O	X	O
O	X	O	X	O		O
O		O				O
						O

6. Upside Breakout above a Bullish Triangle

		X				
		X	O			X
		X	O	X		X
		X	O	X	O	X
		X	O	X	O	X
		X	O	X	O	
		X	O	X		
		X	O			
		X				
		X				
X		X				
X	O	X				
X	O	X				
	O					

6. Downside Breakout of a Bearish Triangle

	X					
O	X	O				
O	X	O				
O	X	O				
O		O				
		O				
		O	X			
		O	X	O		
		O	X	O	X	
		O	X	O	X	O
		O	X	O	X	O
		O	X	O		O
		O	X			O
		O				

FIGURE C.4 Essential Point-and-Figure Chart Patterns

7. Upside Breakout above a Bullish Resistance Line

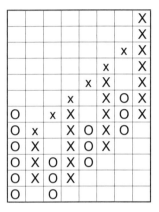

7. Downside Breakout Below a Bullish Support Line

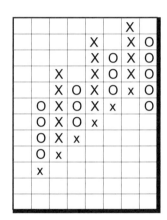

8. Upside Breakout above a Bearish Resistance Line

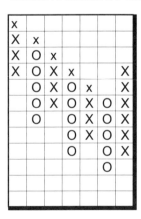

8. Downside Breakout Below a Bearish Support Line

FIGURE C.4 Essential Point-and-Figure Chart Patterns

Appendix C — The Active/Passive Indicator

THE BASIC BUY SIGNAL

When a column of X's rising above the previous column of X's, we have a basic buy signal (see Table C.4). Of the 16 individual patterns, this is the most frequent.

TABLE C.4
The Basic Buy Signal

	1	2	3
64.40			
64.00			X
63.60	X		X
63.20	X	O	X
62.80	X	O	X
62.40	X	O	
62.00	X		

A basic sell signal occurs when a column of O's moves below a previous column of O's (Table C.5).

TABLE C.5
The Basic Sell Signal

	1	2	3	4
20.00				
19.80	O	7		
19.60	O	X	O	
19.40	O	X	O	
19.20	6		O	
19.00			O	
18.80				

The next two basic patterns, the basic buy with a rising bottom (Table C.6), and the basic sell with a declining top (Table C.7), occur more often than any other point-and-figure formation. Both patterns require a base of four columns.

COMPLEX BUY SIGNALS

Triple Top Breakout

The triple top breakout has the highest ratio of profitability for any buy formation — 87.9 percent of the time this pattern gives a successful signal. The last column of X's must move one box above the previous column of X's. The triple top breakout unfolds over a base of five columns.

TABLE C.6
The Basic Buy with Rising Bottom

10.80					
10.60				X	
10.40		X		X	
10.20	O	X	O	X	
10.00	O	X	O	X	
9.90	O	X	O		
9.80	O				
9.70					

TABLE C.7
The Basic Sell with Declining Top

10.70				
10.60	X			
10.50	X	O	X	
10.40	X	O	X	O
10.30		O	X	O
10.20		O		O
10.00				O

TABLE C.8
Triple Top Breakout

5.70					
5.60					X
5.50	X		X	O	X
5.40	X	O	X	O	X
5.30	X	O	X	O	X
5.20		O			X
5.10					

Triple Bottom Breakout

The mirror image of this pattern is the triple bottom breakout (Table C.9). The triple bottom breakout has an even higher profitability ratio than its opposite — with an accuracy rate of 93.5 percent. This pattern is an extremely important sell signal.

Appendix C — The Active/Passive Indicator

TABLE C.9
Triple Bottom Breakout

6.80						
6.70		X		X		
6.60	O	X	O	X	O	
6.50	O	X	O	X	O	
6.40	O				O	
6.30					O	
6.20						

Rising Triple Top

The rising triple top formation (Table C.10) also takes five columns to form. It is stronger than the breakout of a triple top because the pattern contains two basic buy signals, and both X and O columns are rising with the formation.

TABLE C.10
Rising Triple Top

22.60						
22.40					X	
22.20			X		X	
22.00	X		X	O	X	
21.80	X	O	X	O	X	
21.60	X	O	X	O		
21.40		O				

Falling Triple Bottom

The falling triple bottom sell pattern (Table C.11) is the exact opposite of the rising triple top buy pattern. The pattern also unfolds over five columns, contains two basic sell signals, and each X and O column is lower than the previous.

TABLE C.11
Descending Triple Bottom

25.00						
24.80		X				
24.60	O	X	O	X		
24.40	O	X	O	X	O	
24.20	O		O	X	O	
24.00			O		O	
23.80					O	

Spread Triple Top

The spread triple top buy formation (Table C.12) takes seven columns to form and has a profitability percent of 85.7 percent.

TABLE C.12
Spread Triple Top

	1	2	3	4	5	6	7
16.00							
15.80							X
15.60	X		X				X
15.40	X	O	X	O	X		X
15.20	X	O	X	O	X	O	X
15.00		O		O	X	O	X
14.80				O		O	
14.60							

The mirror image of the spread triple top is the spread triple bottom (Table C.13). This is a seven-column pattern with an accuracy rate of 86.5 percent.

TABLE C.13
Spread Triple Bottom

	1	2	3	4	5	6	7
16.40	O						
16.20	O	X		X			
16.00	O	X	O	X	O	X	
15.80	O	X	O	X	O	X	O
15.60	O	X	O	X	O	X	O
15.40	O	X	O	X	O		O
15.20	O		O				O
15.00							O

Bullish Triangle

The bullish triangle (Table C.14) can be a powerful sling that propels an ETF's price considerably higher. It gives accurate signals 71.4 percent of the time. To qualify as a triangle, the formation must develop over five columns.

Bearish Triangle

The opposite of the bullish triangle is the bearish triangle (Table C.15) It also features rising bottoms and declining tops over a five-column base. The bears win control with a basic sell signal on the far right of the chart. This formation had an accuracy rating of 87.5 percent.

Appendix C — The Active/Passive Indicator

TABLE C.14
Bullish Triangle

28.60						
28.40	X				X	
28.20	X	O			X	
28.00	X	O	X		X	
27.80	X	O	X	O	X	
27.60	X	O	X	O	X	
27.40	X	O	X	O		
27.20	X	O	X			
27.00	X	O				
26.80	X					
26.60	X					
26.40						
26.20						
26.00						

TABLE C.15
Bearish Triangle

46.80						
46.40	O					
46.00	O					
45.60	O	X				
45.20	O	X	O			
44.80	O	X	O	X		
44.40	O	X	O	X	O	
44.00	O	X	O	X	O	
43.60	O	X	O		O	
43.20	O	X			O	
42.80	O				O	
42.40						
42.00						
41.60						

TRENDLINES

Prices move up and down over time, occasionally breaking out to move in a consistent direction. This is the beginning of a trend. Once a trend begins, it usually persists for some time. In P&F charting, the upper and lower boundaries of trends are marked with trendlines. P&F trendlines are drawn in only two directions, either diagonally up to the right or diagonally down to the left (Table C.16). They automatically extend out at the proper angle. Draw them by placing either a small case x in each box, or use a solid line.

TABLE C.16
P&F Trendlines

46.80							
46.40					X		
46.00				X			
45.60			X				
45.20		X					
44.80	X						
44.40	X						
44.00		X					
43.60			X				
43.20				X			
42.80					X		
42.40						X	
42.00							
41.60							

Note: 44.40 has X in col 1; 44.80 has X in col 2.

Bullish Support Line

Bullish support lines support prices as they trend up. Prices are in an uptrend when each rally reaches a higher high and each decline reverses at a higher level. Uptrends can last for days, weeks, or several years. In P&F charts, uptrend lines are drawn at 45-degree angles starting at the lowest point to the lower left of the graph where the uptrend begins. It follows along just underneath the trend like a flight of stairs. In the next chart an uptrend begins at $25.00 (Table C.17). Each rally rises a little higher before falling back to touch the stairs. Each bottom is slightly higher than the previous. This is the bullish support line.

TABLE C.17
Bullish Support Line

27.40									
27.20						X			
27.00	O					X	O		
26.80	O			X		X	O		
26.60	O	X		X	O	X	O		x
26.40	O	X	O	X	O	X		x	
26.20	O	X	O	X	O	X	x		
26.00	O	X	O	X	O	x			
25.80	O	X	O	X	x				
25.60	O	X	O	x					
25.40	O	X	x						
25.20	O	x							
25.00	x								
24.80									

Appendix C — The Active/Passive Indicator

Bullish support lines can be drawn after an ETF has formed a trading range below the bearish resistance line and finally gives a buy signal. Begin drawing the line in the box directly under the lowest column of O's in the chart pattern before the buy signal. Connect each box upward in a 45-degree angle.

Bearish Support Line

When prices trend down, they are held up by the bearish support line. In a downtrend, prices fall to lower lows and rallies stop at a lower tops. Downtrends can last for days, weeks, or several years. Support lines are areas of price support. When prices reach areas they bounce back the other way.

Bearish support lines can be drawn anytime there is a column of X's with at least two X's below the lowest O in the next column to the right.

TABLE C.18
Bearish Support Line

Ex. 1

X				
X	O			
X	O	X		
X	O	X		
X	O	X		
X	O			
X	x			
X		x		
			x	
				x

Ex. 2

X				
X	O			
X	O			
X	O			
X	x			
X		x		
X			x	
X				x
X				x

Bullish Resistance Line

Trends tend to move in defined channels. In an uptrend, support is supplied by the bullish support line. Resistance is the bullish resistance line. Bullish resistance lines can be drawn anytime there is a wall of O's with at least two O's above the highest X in the next comumn to the right (see Table C.19).

The ceiling for a downtrend is the bearish resistance line. As prices try to rally back from the decline, they run into this ceiling and fall back to eventually touch the bearish support line.

The bearish resistance line is drawn after an ETF falls back, giving its first sell signal. Begin drawing the trend line in the box directly above the highest X. Connect the boxes diagonally down in a 135-degree angle (Table C.20).

Point-and-figure patterns that involve trendlines are of particular importance. Once an established trendline is broken, the underlying security is changing course and deserves special attention.

TABLE C.19
Bullish Resistance Line

Ex. 1

			X		
		X			
O	X				
O	x				
O	X				
O	X	O			
O	X	O			
O	X	O			
O	X	O			
O	X	O			
O	X				
O					

Ex. 2

				X	
O			x		
O			x		
O		x			
O	x				
O	X				
O	X	O			
O	X	O			
O		O			
O					

TABLE C.20
Bearish Resistance Line

46.80						x			
46.40				X		X	x		
46.00				X	O	X	O	x	
45.60			X	X	O	X	O	X	x
45.20	X	O	X	O	X	O	x	O	X	O
44.80	X	O	X	O	X	x		O	X	O
44.40	X	O	X	O	x			O		O
44.00	X	O	X	x						
43.60	X	O	x							
43.20	X	x								
42.80	x									
42.40										
42.00										
41.60										

Breakout above Bullish Resistance

This is a particularly powerful pattern because it embodies three bullish elements (Table C.21). Prices are in an uptrend, have just issued a basic buy pattern, and have risen above the bullish resistance line. When this occurs, the old resistance line can become the new support line.

Appendix C — The Active/Passive Indicator

TABLE C.21
Upside Breakout above Bullish Resistance Line

25.60									X
25.40								X	x
25.20								X	
25.00							x	X	
24.80						x		X	
24.60					x	X		X	
24.40				x		X	O	X	
24.20			x	X		X	O	X	
24.00	O	x		X	O	X	O		
23.80	O	X		X	O	X			
23.60	O	X	O	X	O				
23.40	O	X	O	X					
23.20	O		O						
23.00									

Downside Breakout below Bullish Support Line

It is bad news when prices fall below a bullish support line as it often signals the end of an uptrend, or at least a fairly lengthy interruption in the longer-term trend. Once support or resistance lines are crossed, it means that momentum has changed and ETF sellers now outnumber buyers.

TABLE C.22
Downside Breakout Below Bullish Support Line

24.60								
24.40								
24.20								
24.00								
23.80						X		
23.60					X	X	O	
23.40					X	O	X	O	x
23.20			X		X	O	X	O	
23.00	X		X	O	X	O	x	O	
22.80	X	O	X	O	X	x		O	
22.60	X	O	X	O	x				
22.40	X	O	X	x					
22.20	X	O	x						
22.00	X	x							

Breakout above a Bearish Resistance Line

Here the price trend is down, below the bearish resistance line, when it breaks out above the line. This often marks the beginning of a new trend and allows you to buy the emerging up-trend about as near the bottom as you can get.

TABLE C.23
Breakout Above Bullish Resistance Line

24.20	X							
24.00	X	X						
23.80	X	O	x					
23.60	X	O	X	x		X		
23.40		O	X	O	x	X		
23.20		O	X	O	X	x	X	
23.00		O		O	X	O	X	
22.80				O	X	O	X	x
22.60				O			O	X
22.40							O	
22.20								
22.00								

Downside Breakout below Bearish Support Line

The breakout below a bearish support line is very negative (Table C.24). The ETF is descending but then it breaks through support and falls rapidly.

TABLE C.24
Downside Breakout Below Bearish Support Line

24.20								
24.00		X						
23.80	O	X	O					
23.60	O	X	O	X				
23.40	O		O	X	O			
23.20	x		O	X	O	X		
23.00		x	O		O	X	O	
22.80			x		O	X	O	
22.60				x	O		O	
22.40					x		O	
22.20						x	O	
22.00							O	
21.80							O	x
21.60								

Appendix C — The Active/Passive Indicator

ESTIMATING PRICE OBJECTIVES

Point-and-figure charts are remarkably effective for forecasting price objectives. The ability to predict percentage price moves allows the investor to estimate whether an ETF is worth buying or not.

There are two ways to calculate price objectives, the horizontal count and the vertical count. The vertical count is usually more reliable, but be familiar with both.

Vertical Count

Use the vertical count (Table C.25) to compute price objectives after a buy or sell signal. If you are on a buy signal, determine the upside target using the vertical count by first moving back to the left to find the lowest point after the most recent *sell* signal. From that low point, count the number of X's in the next column to the right. Multiply that result by the box size and then multiply again by three. Add the result to the low point after the most recent sell signal.

A basic buy signal is given in the last column at $17.20. Find the lowest point on the chart just after the prior *sell* signal. The most recent sell signal was given at $16.20 and the lowest point since that signal occurred at $14.40. Count the number of X's just to the right of the low point and multiply by 3 (because we are using a

---

TABLE C.25
Vertical Count

17.80								
17.60			X					
17.40			A	O				
17.20			X	O		X		
17.00			X	O	C	X		
16.80	X		X	O	X	O	X	
16.60	X	O	X	O	X	O	3	
16.40	7	O	X	B		O	X	
16.20	X	8	X			O	X	
16.00	X	O	X			O	X	
15.80	X	9				O	X	
15.60	6					O	X	
15.40	X					O	2	
15.20	X					O	X	
15.00	X					O	1	
14.80	X					O	X	
14.60	5					O	X	
14.40	3					O		
14.20	2							
		9						
		9						

TABLE C.26
Changing Box Value

Price						
44.40						
44.00						
43.60	C					
43.20	O	2				
42.80	O	X	O			
42.40	O	X	O			
42.00	O	X	O			
41.60	O	X	3			
41.20	O	1	O			
40.80	O	X	O			
40.40	O	X	O			X
40.00	O	X	O	X		X
39.80	O	X	O	X	O	X
39.60	O		O	X	O	X
39.40			O	X	O	X
39.20			O	X	O	X
39.00			4		O	X
38.80					O	
	9 9		9 9			

three-box reversal). There are 14 boxes filled with X's (14 × 3 = 42). At this price level on the chart, each box is worth 20¢ (42 × 20¢ = $8.40). Add this to the low of $14.40 and get a price objective of $22.80.

The price could go higher or not. Often an ETF will hit the price objective and then continue higher. Always place greater emphasis on trendlines and relative strength. Just because an ETF meets your price objective doesn't mean you have to sell. It does prompt you to reevaluate from that level.

In the next example (Table C.26), the box values change from 20¢ to 40¢ per box right in the first column of X's off the bottom.

The first buy signal off the bottom is at $40.40. The low after the prior sell signal is at $38.80. Move to the right and count the X's up to the buy signal. That is seven X's. Be sure to notice that the first six X's are in 20¢ boxes while the last one is in a 40¢ box. Multiply 20¢ times 6, and 40¢ times 1, then add the results together.

$$0.2 \times 6 = 1.20$$

$$0.40 \times 1 = 0.40$$

$$1.20 + 0.40 = 1.60$$

Appendix C — The Active/Passive Indicator

TABLE C.27
Downside Price Objective

27.20							
27.00				X			
26.80				X	O		
26.60				X	O	X	
26.40	X			X	O	X	O
26.20	X	O	2	O	5	O	
26.00	X	O	X	3	X	O	
25.80	X	O	1	O	X	O	
25.60	X	C	X	O		O	
25.40	B	O	X			O	
25.20	X	O					
25.00	X						
24.80	A						
24.60	9						
24.40							
	9/5			9/6			

Now multiply $1.40 times three and add the result to the low price of $39.20.

$$1.60 \times 3 = 4.80$$

$$38.80 + 4.80 = 43.60$$

The upside target price is $43.60.

Calculating downside price objectives using the vertical count is merely the reverse as the upside, with one slight difference. Instead of multiplying by 3, multiply by 2. In Table C.27, a basic sell signal is given at $25.40. Move back to the left until you reach the highest point after the most recent *buy* signal at $26.60. Count the number of O's in the very next column. Multiply that number, which in this case is seven, by the current box size of 20¢, and then multiply that result by 2. Subtract that number from the highest point at $27.00. The outcome is the downside target.

Box size of 20¢:

$$0.20 \times 7 \text{ (seven O's occur after the highest point)} = 1.40$$

Multiply result from above by 2:

$$1.40 \times 2 = 2.80$$

TABLE C.28
Double Top with Changing Box Value

Price							
42.80							
42.40			X		X		
42.00			X		X	O	
41.60			X	O	2	O	
41.20	X		X	O	X	3	
40.80	X	O	9	B	1	O	
40.40	X	O	X	O	X	O	
40.00	X	O		C		O	X
39.80	X					O	X
39.60	X					O	X
39.40	X					4	X
39.20						O	
39.00							
38.80							
38.60							
38.40							
38.20							
			9 7		9 8		

Subtract the above result from the value of the highest point:

$$27.00 - 2.80 = 24.20$$

The downside price target is $24.20.

The first sell signal is at $39.80. Go back to the high after the prior *buy* signal. The prior buy was at $41.60 and the top was a double top at $42.40. In the case of a double top, use the latest price. Move one column to the right and count the O's down to the sell signal. That's ten O's. Notice that the first six O's are in 40¢ boxes while the last four are in 20¢ boxes. Therefore you have to multiply 40¢ times 6, and 20¢ times 4, then add the results together.

$$0.40 * 6 = 2.40$$

$$0.20 \times 4 = 0.80 \quad 2.40 + 0.80 = 3.20$$

Now multiply $3.20 times <u>two</u> and subtract the result to the high price of $42.40.

$$3.20 \times 2 = 6.40$$

$$42.40 - 3.20 = 39.20$$

The downside price objective is $39.20.

Horizontal Measurement

The horizontal count (Table C.29) also involves counting boxes across either the base or the top of a pattern. Use the horizontal count to determine price objectives whenever the chart gives buy signals from a bottom or sell signals from a top pattern. Count the number of boxes across a pattern, multiply by three and then by the value of the boxes. Add or subtract that value from the low or the high to arrive at the target.

TABLE C.29
Horizontal Count

16.00						
15.80						X
15.60		X		8		X
15.40	O	5	O	7	O	B
15.20	O	X	O	X	O	X
15.00	4	X	6	X	9	X
14.80	O			O		A
14.60						
	9				9	
	8				9	

After a seven-month trading range, the chart gives a buy signal in November 1994, at $15.80. In a trading range or basing pattern, prices go up, hit a ceiling, fall back to hit a floor, and rebound upward again. More sellers who wanted out of the ETF at $15.60 while more buyers wanted to purchase shares at $14.80.

To predict how high the rally could carry using the horizontal count, do the following. Up to the buy signal, prices moved in a trading range, or basing pattern, six boxes wide. We are using three-box reversal charts, so we multiply six times three.

$$6 \times 3 = 18$$

At this level on the price chart, each box is worth 20¢, so multiply 18 times 20¢.

$$18 \times 0.20 = 3.60$$

Finally, add this to the lowest point in the basing pattern, which is $14.20, and get the upward price target.

$$\$14.80 + 3.60 = \$18.40$$

For downside targets merely subtract from the high. Observe the next example (Table C.30).

A three-box reversal is worth 60¢ at this price level on the chart.

$$\$0.20 \times 3 = \$0.60$$

TABLE C.30
Downside Target using Horizontal Count

	1	2	3	4	5	6	7	8
40.80								
40.40	X		X		12			
40.00	X	O	X	O	X	O		
38.80	X	O	X	O	X	O		
38.60	X	O			O	X	O	
38.40	11				O	X	O	
38.20	X				O		O	
38.00	X						O	
37.80								
37.60								
	9						9	
	6						7	

The basing pattern before the sell signal at $38.00 is six boxes wide. Multiply the six boxes times the value of a three-box reversal.

$$\$0.60 \times 6 = \$3.60$$

Subtract $3.60 from the top at $40.40 to arrive at our price objective.

$$\$40.40 - 3.60 = \$36.80$$

Determining price targets comes into play when determining risk-reward relationships. Demand twice the level of potential profit over loss before making any purchase. For instance, you find an ETF on a new price buy signal, positive relative strength, and above its trendline. Calculate a vertical or horizontal count to derive an upside target. If the target does not afford twice the possible gain as the loss we would occur if the ETF fell back below the trendline on a sell signal, don't buy the ETF.

OTHER HIGH PROBABILITY COMPLEX PATTERNS

High Pole Formation

A high pole formation occurs when a point-and-figure chart rises above a previous column of X's by at least three boxes only to reverse back into a column of O's and retrace at least 50 percent of the move (Table C.31). The pattern shows a problem is forming with the supply/demand relationship and that supply might be taking over control. Be suspicious and ready to exit.

Low-Pole Formation

A low-pole formation occurs when a column of O's falls at least three boxes below the previous column of O's and then retraces at least 50 percent (Table C.32).

TABLE C.31
High Pole Sell Formation

	1	2	3	4
13.00				
12.80				
12.60				
12.40			X	
12.20			X	O
12.00			X	O
11.80			X	O
11.60			X	O
11.40	X		X	O
11.20	X	O	X	
11.00	X	O	X	
10.80	X	O		
10.60	X			
10.40	X			
10.20				

TABLE C.32
Low Pole Buy Formation

	1	2	3	4	5
13.00					
12.80	O				
12.60	O	X			
12.40	O	X	O		
12.20	O	X	O		
12.00	O	X	O	X	
11.80	O		O	X	?
11.60			O	X	?
11.40			O	X	?
11.20			O	X	
11.00			O		
10.80					
10.60					
10.40					
10.20					

Multiple Box Up Formation

Sometimes an ETF will be so hot that its price will rise straight up 20 or more boxes without a break. It has to be 20 or more to qualify as a multiple box up formation (Table C.33). Review this formation in the text. You run into it quite often in bull markets and will want to know what to do. A good point to take some profits is the first three-box reversal.

TABLE C.33
Multiple Box Up Formation

	95	96		97	
16.40					
16.20					
16.00					
15.80					
15.60					
15.40		X			
15.20		X	O		
15.00		11	O		
14.80		X	O	——— Good point to take some profits.	
14.60		10			
14.40		X			
14.20		X			
14.00		X			
13.60		X			
13.40		X			
13.20		X			
13.00		X			
12.80		X			
12.60		X			
12.40		X			
12.20		X			
12.00		X			
11.80		X			
11.60		X			
11.40		X			
11.20		X			
11.00		X			
10.80		1			
10.60		X			
10.40		X			
10.20		X			
10.00	O	X			
9.90	O	X			
9.80	O	X			
9.70	O	X			
9.60	O				
9.50					
9.40					
9.30					

On a relative strength chart, or if this pattern occurs on the passive/active indicator, the price of the diversified core would have to virtually collapse

Appendix C — The Active/Passive Indicator

relative to the S&P 500 in order for a traditional sell signal to be given. Rather than wait for the column of O's to fall below a previous O column, it is best to call the first three-box reversal off the top negative until a definitive pattern can form. This is the only time to act without a definitive signal, but is extremely prudent.

Multiple Box Down Formation

When a chart has fallen over 20 boxes without interruption, the first three-box reversal into a column of X's usually presents a good buying opportunity (Table C.34). However, the longer it takes for the ETF to descend 20 boxes, the less dependable the pattern is.

TABLE C.34
Multiple Box Down Formation

	95	96	97		
14.00					
13.60	O				
13.40	O	X			
13.20	O	X	O		
13.00	O	X	O		
12.80		X	O		
12.60			O		
12.40			O		
12.20			O		
12.00			O		
11.80			O		
11.60			O		
11.40			O		
11.20			O		
11.00			O		
10.80			O		
10.60			O		
10.40			O		
10.20			O		
10.00			O		
9.90			O	X	← Possible entry point for purchase.
9.80			O	X	
9.70			O	X	
9.60			O		
9.50					
9.40					
9.30					

Broadening Top Formation

A broadening top formation occurs after an ETF has made a good upmove, and then reaches a high (1), followed by a low (2), then a higher top (3), followed by a lower low (4), and finally a higher high (5) over at least five columns (Table C.35). This is a bearish formation and signals a substantial correction may follow.

TABLE C.35
Broadening Top Chart

14.20								
14.00							(5)	
13.80							X	
13.60							X	
13.40						(3)	X	
13.20		(1)				X	X	
13.00		X				X	O	X
12.80	O	X	O	X		X	O	X
12.60	O	X	O	X	O	X	O	X
12.40	O	X	O	X	O	X	O	X
12.20	O		O		O		O	X
12.00			O				O	X
11.80		(2)					O	
11.60						(4)		
11.40								
		9					9	
		6					7	

Index

A

Active account analysis process, 84
Active management, 69–70
Active/passive
  indicator, 74, 227
  investment vehicles, 83
American Stock Exchange (AMEX), 13–14
  creation of basket securities and, 7
  equity index participations, 6
  website, 14
Asia/Pacific Region funds, 118–119, 180–186
Asset allocation, 21
Australian Index Fund, 180
Austria Index Fund, 190

B

B2B Internet HOLDRS, 206
Barclays Global Investors, N.A. (BGI), 14–16
  fixed-income baskets, 111
Basic buy signal, 235
Basic Industries Select Sector SPDR Fund, 155
Basic industry stocks, 97, 118, 154–157
Basis, 5
Basket case portfolio, 30, 71
  active portion of, 83
  active/passive indicator, 227
  value at risk for, 75
Basket securities
  benefits of, 11–12
  broad-based index, 13
  Country Baskets, 16
  definition of, 9
  development of, 3
  exchange trading, 10
  fixed income, 111–112
  HOLDRS, 16–17
  iShares, 15–16, 107
  large cap, 129–148
  mid cap, 125–128
  momentum strategy for international equity markets, 110
  portfolio deposits and redemptions, 10
  small cap, 120–124
  structure of, 9
  trading strategies for, 113
  use of in equity portfolios, 21–27
Basket trading, 5
Bearish
  resistance line, breakout above, 244
  support line, 241
    downside breakout below, 244
  triangle, 238
Belgium Index Fund, 189
Biotech HOLDRS, 204
Bloomberg, 71
Bombshells, reduction of, 11–12
Bond indexes, exchange-traded funds based on, 111–112
Brazil Index Fund, 201
Broadband HOLDRS, 205
Bullish
  resistance line, 241
    breakout above, 242
  support line, 240
    downside breakout below, 243
  triangle, 238
Business cycle
  definition of, 84
  five phases of, 88
  indicators of, 85–86
  sector analysis, 92–100
  tabular representation of, 89
  technical analysis of, 100–103
Buy signal
  basic, 235
  complex, 235–254
Buy-and-hold strategy, 110

C

Canada Index Fund, 188
Canadian funds, 119, 187–188
Cap-weighted equal-weighted index comparison, 71
Capital goods, 95–96
Capitalization-weighted index, 71
Cash index participations, 6
Charles Schwab, 112
Chemicals, 118, 162
Chicago Board Options Exchange, 6

255

Chicago Mercantile Exchange, 6
Cohen & Steers Realty Major Index Fund, 177
Coincident indicators, 85
    industrial production, 91
Commodity Futures Trading Commission
    (CFTC), 6
Consumer cyclical stocks, 94–95, 98
Consumer expectations, 91
Consumer Price Index (CPI), 91
Consumer services, 96
Consumer Services Select Sector SPDR Fund,
    163
Consumer staples, 98
Consumer Staples Select Sector SPDR Fund, 164
Consumer stocks, 118, 163–164
Cost efficiency, 11
Cost of carry, 5
Country Baskets, 16, 107
    momentum strategy approach to, 110
Crash, October 1987, 3
Creation unit transactions, 10
Currency movement, 109
Cyclical stocks, 118, 174–176
Cyclical/Transportation Select Sector SPDR
    Fund, 174

D

Derivative instrument, 4
Designated Order Turnaround, 4
DIAMONDS, 13, 16, 142
Discount rate, 92
Diversification, 21
    international, 108–110
    thousand stock portfolio, 22–27
Diversified core, 22, 69
    companies used in account for Thousand
        Stock Portfolio strategy, 31–54
    determining the allocation of, 72–74
    determining the blend, 74–81
    developing value at risk for, 75–78
    semiactive, 27
    semipassive, 29
Dow Jones
    Global Titans Index Fund, 138
    Industrial Average, 142
    U.S. Basic Materials Sector Index Fund, 154
    U.S. Chemical Sector Index Fund, 162
    U.S. Consumer Cyclical Sector Index Fund,
        175
    U.S. Consumer Non-Cyclical Sector Index
        Fund, 176
    U.S. Energy Sector Index Fund, 159
    U.S. Financial Sector Index Fund, 150
    U.S. Financial Services Sector Index Fund,
        151
    U.S. Healthcare Sector Index Fund, 153
    U.S. Industrial Sector Index Fund, 157
    U.S. Internet Sector Index Fund, 172
    U.S. Large Cap Growth Index Fund, 131
    U.S. Large Cap Value Index Fund, 141
    U.S. Real Estate Index Fund, 178
    U.S. Small Cap Growth Index Fund, 123
    U.S. Small Cap Value Index Fund, 124
    U.S. Technology Sector Index Fund, 165
    U.S. Telecommunications Sector Index, 173
    U.S. Total Market Index Fund, 139
    U.S. Utilities Sector Index Fund, 161
Dynamic Asset Allocation (DAA), 6

E

EAFE index, 107
Early contraction, 97–98
Early expansion, 94–95
Economic contraction, 97–100
Economic expansion, 94–97
Economic indicators, 85
Efficient market hypothesis, 101
Emerging markets, 109
Energy Select Sector SPDR Fund, 158
Energy stocks, 96–97, 118, 158–161
Equal-weighted index, 71
Equitization of cash, 69–70
Equity exposure, 3
Equity index participations, 6
Equity portfolios
    components of, 69
    internationally diversified, 107–110
    use of basket securities in, 21–27
Europe 2001 HOLDRS, 207
European funds, 119, 189–200
European Monetary Union (EMU) Index Fund,
    193
Excess returns, positive risk-adjusted, 69
Exchange trading, 10
Exchange-traded funds. *See also* basket securities
    AMEX website for, 14
    based on U.S. government bond indexes,
        111–112

F

Falling triple bottom, 237
Federal Funds, 91
Fidelity Investments, 112
Financial Select Sector SPDR Fund, 149

Index

Financials, 99, 117–118, 149–151
Five-stage system, 67
Fixed Income Trust Receipts, 111
FOLIOfn Inc., 112
Fortune 500 Index Tracking Stock, 135
Fortune e-50 Index Tracking Stock (FEF), 14, 169
France Index Fund, 198
Freshness, 11
Futures, 3
    index arbitrage and, 5

G

Germany Index Fund, 192
Goldman Sachs Technology Index Fund, 168

H

Healthcare stocks, 98, 118, 152–153
Hedging strategies, 113
Historical method to compute value at risk, 75
Holding Company Depositary Receipts. *See* HOLDRS
HOLDRS, 16–17
    B2B Internet, 206
    biotech, 204
    broadband, 205
    description of, 203
    Europe 2001, 207
    internet, 208
    internet architecture, 209
    internet infrastructure, 210
    Market 2000+, 211
    oil service, 212
    pharmaceutical, 213
    regional bank, 214
    retail, 215
    semiconductor, 216
    software, 217
    telecom, 218
    utilities, 219
    wireless, 220
Hong Kong Index Fund, 183

I

Index arbitrage, 3
    description of, 4–5
Index participations, 6
Index Trust SuperUnit, 6, 14
Indicative Optimized Portfolio Value (IOPV), 10
Individual active account sector, companies in, 55–67
Industrial production, 91
Industrial Select Sector SPDR Fund, 156
Inflation, 91
Instant diversification, 9
Interest rates, 91
International investing
    momentum strategies for, 110
    reassessing, 108–110
Internet Architecture HOLDRS, 209
Internet HOLDRS, 208
Internet Infrastructure HOLDRS, 210
Investment Company Act of 1940, 6
Investment tools, 68
iShares
    bond indexes, 111
    Cohen & Steers Realty Majors Index Fund, 177
    Dow Jones U.S. Basic Materials Sector Index Fund, 154
    Dow Jones U.S. Chemical Sector Index Fund, 162
    Dow Jones U.S. Consumer Cyclical Sector Index Fund, 175
    Dow Jones U.S. Consumer Non-Cyclical Sector Index Fund, 176
    Dow Jones U.S. Energy Sector Index Fund, 159
    Dow Jones U.S. Financial Sector Index Fund, 150
    Dow Jones U.S. Financial Services Sector Index Fund, 151
    Dow Jones U.S. Healthcare Sector Index Fund, 153
    Dow Jones U.S. Industrial Sector Index Fund, 157
    Dow Jones U.S. Internet Sector Index Fund, 172
    Dow Jones U.S. Real Estate Index Fund, 178
    Dow Jones U.S. Technology Sector Index Fund, 165
    Dow Jones U.S. Telecommunications Sector Index, 173
    Dow Jones U.S. Total Market Index Fund, 139
    Dow Jones U.S. Utilities Sector Index Fund, 161
    Goldman Sachs Technology Index Fund, 168
    MSCI Series, 14, 107
        Australia Index Fund, 180
        Austria Index Fund, 190
        Belgium Index Fund, 189
        Brazil Index Fund, 201
        Canada Index Fund, 188
        EMU Index Fund, 193
        France Index Fund, 198
        Germany Index Fund, 192
        Hong Kong Index Fund, 183

Italy Index Fund, 195
Japan Index Fund, 185
Malaysia (Free) Index Fund, 186
Mexico Index Fund, 202
Netherlands Index Fund, 196
Singapore Index Fund, 182
South Korea Index Fund, 184
Spain Index Fund, 191
Sweden Index Fund, 200
Switzerland Index Fund, 197
Taiwan Index Fund, 181
United Kingdom Index Fund, 199
use of in a Thousand Stock Portfolio strategy, 30
Nasdaq Biotechnology Index Fund, 152
Russell 1000 Growth Index Fund, 132
Russell 1000 Index Fund, 140
Russell 1000 Value Index Fund, 145
Russell 2000 Growth Index Fund, 146
Russell 2000 Index Fund, 147
Russell 2000 Value Index Fund, 148
Russell 3000 Growth Index Fund, 130
Russell 3000 Index Fund, 136
Russell 3000 Value Index Fund, 144
S&P 500 Index Fund, 137
S&P 500/BARRA Growth Index Fund, 133
S&P 500/BARRA Value Index Fund, 143
S&P Europe 350 Index Fund, 194
S&P MidCap 400 Index Fund, 126
S&P MidCap 400/BARRA Growth Index Fund, 125
S&P MidCap 400/BARRA Value Index Fund, 128
S&P small cap 600 index fund, 120
S&P small cap 600/BARRA Growth Index Fund, 121
S&P small cap 600/BARRA Value Index Fund, 122
S&P/TSE 60 Index Fund, 187
Italy Index Fund, 195

J

Japan Index Fund, 185

L

Lagging indicators, 85
Large cap funds, 117, 129–148
Late contraction, 99–100
Late expansion, 96–97
Latin American funds, 119, 201–202
Leading indicators, 85
  consumer expectations, 91

stock market as a, 87
stock prices as, 92
Leland O'Brien Rubinstein Associates Incorporated, 6, 221
Liquidity, 10

M

Malaysia (Free) Index Fund, 186
Margin requirements, 92
Market 2000+ HOLDRS, 211
Market baskets, 5
Market exposure, 3
Merrill Lynch, 16–17
Merrill Lynch, Pierce, Fenner & Smith Incorporated. *See* Merrill Lynch
Mexico Index Fund, 202
Mid cap funds, 117, 125–128
Midcap SPDR, 13
  use of in Thousand Stock Portfolio strategy, 30
MidCap SPDR Trust Series I, 127
Middle expansion, 95–96
Model Portfolio, 71
Momentum strategy, 110
Money Market Trust SuperUnit, 7
Monte Carlo Simulation, 75
Morgan Stanley
  Capital International 20-country EAFE index, 107
  Capital International Inc. (MSCI), 14
  High Tech 35 Index Fund, 167
  High Tech 35 Index Fund (MTK), 14
  Internet Index Fund (MII), 14, 171
MSCI
  Australia Index Fund, 180
  Austria Index Fund, 190
  Belgium Index Fund, 189
  Brazil Index Fund, 201
  Canada Index Fund, 188
  EMU Index Fund, 193
  France Index Fund, 198
  Germany Index Fund, 192
  Hong Kong Index Fund, 183
  Italy Index Fund, 195
  Japan Index Fund, 185
  Malaysia (Free) Index Fund, 186
  Mexico Index Fund, 202
  Netherlands Index Fund, 196
  Singapore Index Fund, 182
  South Korea Index Fund, 184
  Spain Index Fund, 191
  Sweden Index Fund, 200
  Switzerland Index Fund, 197
  Taiwan Index Fund, 181
  United Kingdom Index Fund, 199

Index

N

Nasdaq Biotechnology Index Fund, 152
Nasdaq-100 Index Tracking Stock (QQQ), 13, 16, 170
  use of in Thousand Stock Portfolio strategy, 30
National Bureau of Economic Research (NBER), 85
  economic indicators, 86t
Net asset value (NAV), 10
  point-and-figure charts and, 229
Netherlands Index Fund, 196
New York Stock Exchange
  Country Baskets, 16
  Designated Order Turnaround (DOT), 4
Noncyclical stocks, 118, 174–176
Nuveen Investments, 111

O

Oil Service HOLDRS, 212

P

Participating agreement, 10
Passive bond investing, 112
Passive index funds, 29
Passive management, 69–70
PDR Services LLC, 13–14. *See also* American Stock Exchange (AMEX)
Personal funds, 112
Pharmaceutical HOLDRS, 213
Philadelphia Stock Exchange, cash index participations, 6
Point-and-figure charts, 227
  basic signals, 235
  bearish triangle, 238
  bullish triangle, 238
  constructing, 227–229
  essential chart patterns, 230–234
  estimating price objectives, 245–250
  falling triple bottom, 237
  high probability complex patterns of, 250–254
  rising triple top, 237
  spread triple top, 238
  trendlines, 239–244
  triple bottom breakout, 236
  triple top breakout, 235
Portfolios, 3
  construction of, 11
  deposits and redemptions, 10
  insurers, 6
  internationally diversified equity, 107

iShares MSCI, 107
passive, 69
personal fund, 112
sector SPDRs, 83, 107
Thousand Stock, 22–27, 30
uses of basket securities in equity, 21–27
zero-investment, 110
Positive risk-adjusted excess returns, 69
Price efficiency, 10
Price objectives, estimating, 245–250
Producer Price Index (PPI), 91
Program trading, 3
  types of, 4–5

R

Real estate stocks, 118, 177–179
Recession, yield curve as an indicator of, 92
Regional Bank HOLDRS, 214
Relative strength, 72
  charts, 230
  use of computer spreadsheets to calculate, 73
Reserves, 91
Retail HOLDRS, 215
Rising triple top, 237
Russell
  1000 Growth Index Fund, 132
  1000 Index Fund, 140
  1000 Value Index Fund, 145
  2000 Growth Index Fund, 146
  2000 Index Fund, 147
  2000 Value Index Fund, 148
  3000 Growth Index Fund, 130
  3000 Index Fund, 136
  3000 Value Index Fund, 144

S

Sector baskets, 29, 83
Sector rotation, 92–93
Security and Exchange Commission (SEC), 6
Select Sector SPDR Fund, 155
Sell signal, basic, 235
Semiactive diversified core, 27
Semiconductor HOLDRS, 216
Singapore Index Fund, 182
Single-country funds, 107
Small cap funds, 117, 120–124
Software HOLDRS, 217
South Korea Index Fund, 184
Spain Index Fund, 191
SPDR, 7, 13, 16, 21
  Basic Industries Select Sector Fund, 155

characteristics and benefits of, 9–12
Consumer Services Select Sector Fund, 163
Consumer Staples Select Sector Fund, 164
Cyclical/Transportation Select Sector Fund, 174
Energy Select Sector Fund, 158
Financial Select Sector Fund, 149
individual active account sector companies, 55–67
Industrial Select Sector Fund, 156
MidCap Trust Series I, 127
recommended components of account, 67
sector baskets, 14
suitability of as active/passive investment vehicles, 83
Technology Select Sector Fund, 166
Trust Series I, 134
use of in Thousand Stock Portfolio strategy, 30
Utilities Select Sector Fund, 160
Spread triple top, 238
Spreadsheets, use of to calculate relative strength, 73
SPY. See SPDR
Standard & Poor's
   500 Index Fund, 137
   500 Stock Index, 3, 134
      as an economic indicator, 83
   500/BARRA Growth Index Fund, 133
   500/BARRA Value Index Fund, 143
   Depositary Receipt, 134 (See also SPDR)
   Europe 350 Index Fund, 194
   indexes, 72
   MidCap 400 Depositary Receipts, 127
   MidCap 400 Index Fund, 126
   MidCap 400/BARRA Growth Index Fund, 125
   MidCap 400/BARRA Value Index Fund, 128
   Small Cap 600 Index Fund, 120
   Small Cap 600/BARRA Growth Index, 121
   Small Cap 600/BARRA Value Index, 122
   TSE 60 Index Fund, 187
Stock baskets, 112
Stock charting, 227
Stock index futures contracts, 3. See also futures
Stock prices, 92
Stock selection, 29
Straight program trading, 4
Strategic asset allocation, 69
streetTRACKS
   Dow Jones Global Titans Index Fund, 138
   Dow Jones U.S. Large Cap Growth Index Fund, 131
   Dow Jones U.S. Large Cap Value Index Fund, 141
   Dow Jones U.S. Small Cap Growth Index Fund, 123
   Dow Jones U.S. Small Cap Value Index Fund, 124
   Morgan Stanley High Tech 35 Index Fund, 167
   Morgan Stanley Internet Index Fund, 171
   Sector Funds, 14
   Wilshire REIT Index Fund, 179
SuperShare Services Corporation (SSC), 6, 221
SuperTrust
   actual outcomes of, 225
   development of, 6–7
   payoffs of, 222–224
   securities of, 222
   structure of, 221
Sweden Index Fund, 200
Sweep account, 69
Switzerland Index Fund, 197

T

Taiwan Index Fund, 181
Tax efficiency, 10
Technology Select Sector SPDR Fund, 166
Technology stocks, 94, 118, 165–173
Telecom HOLDRS, 218
Thousand stock portfolio, 22–27
   strategy for, 30
Toronto Stock Exchange, 187
Trading strategies, 113
Transportation stocks, 94
Triple bottom breakout, 236
Triple top breakout, 235

U

U.S. government bond indexes, exchange-traded funds based on, 111–112
Unit investment trusts (UITs), 9
   cost efficiency of, 11
United Kingdom Index Fund, 199
Utilities, 99–100, 118, 158–161
   HOLDRS, 219
Utilities Select Sector SPDR Fund, 160

V

Value at risk (VAR), 69
   use of to determine diversified core blend, 74–81
Vanguard Total Stock Market VIPERs, 129

Variance/Covariance method to compute value at risk, 75
Volatility of stocks, 70

W

WEBS, 14. *See also* iShares, MSCI Series
Wells Fargo Nikko Investment Advisors, 14. *See also* Barclays Global Investors, N.A. (BGI)
Wilshire 5000 Total Market Index, 129
Wilshire REIT Index Fund, 179

Wireless HOLDRS, 220
World Equity Benchmark Shares. *See* WEBS

Y

Yield curve, 92

Z

Zero-investment portfolios, 110